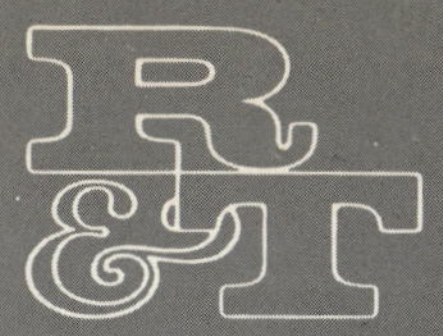

ROAD&TRACK
ON
PORSCHE
1979-1982

Reprinted From
Road & Track Magazine

ISBN 0 907 073 786

Published By
Brooklands Books with permission of Road & Track

Printed in Hong Kong

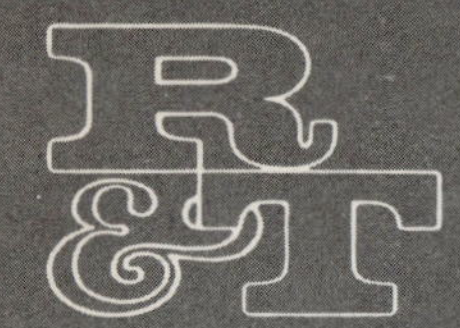

Porsche Titles in this Series

ROAD & TRACK ON PORSCHE 1979-1982
Road Tests – New Model Reports – Owner Surveys – Used Car Classic Reports
Model Updates – Comparison Tests – History – Profile – 4WD 911 – 911SC
914 – 914/6 – 917 – 924 – 924S – 924S Turbo – 924 Supercharged – 928 – 944

ROAD & TRACK ON PORSCHE 1975-1978
Road Tests – New Model Reports – Works Visits – Driving Impressions
Used Car Classic Report – Technical Analysis – Salon – Comparison Tests
904 GTS Carrera – 911S – 911SC – Turbo Carrera – 914 2.0 – 917K
924 – 928 – Speedster Replica

Other titles in this series
Road & Track on Corvette 1953-1967
Road & Track on Corvette 1968-1982
Road & Track on Ferrari   1968-1974
Road & Track on Ferrari   1975-1981
Road & Track on Fiat Sports Cars 1968-1981
Road & Track on Mercedes Sports & GT Cars 1970-1980

Titles in preparation will cover:–
Lamborghini, Jaguar, BMW, Alfa Romeo etc.

Distributed By

Road & Track
1499 Monrovia,
Newport Beach,
California 92651, U.S.A.

Brooklands Book Distribution Ltd.,
Holmerise, Seven Hills Road,
Cobham, Surrey KT11 1ES,
England

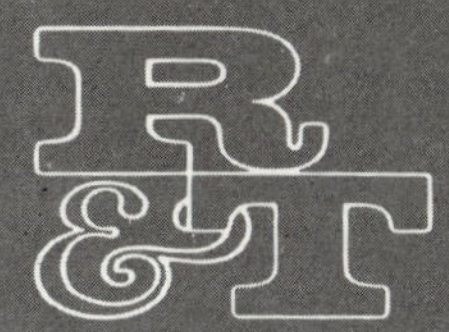

Contents

We are frequently asked for copies of out of print Road Tests and other articles that have appeared in Road & Track. To satisfy this need we are producing a series of books that will include, as nearly as possible, all the important information on one make or subject for a given period.

It is our hope that these collections of articles will give an overview that will be of value to historians, restorers and potential buyers, as well as to present owners of these automobiles.

PORSCHE 924

Evolutionary improvements for Porsche's lowest-price model

PHOTOS BY THOS L. BRYANT & JOE RUSZ

IT HAS ALL the makings of an outstanding GT— 4-wheel independent suspension, single-overhead-camshaft engine with crossflow cylinder head, rear-mounted transaxle, a well designed interior and a stylish exterior. Yet, like a wine that is drunk before maturity, the Porsche 924, now in its third year of production, has yet to exhibit the greatness everyone expected of it. At the time of its introduction early in 1976, the 924 was a minor sensation in the automotive world. It was the first of the new-generation Porsches and as such used a front-mounted, water-cooled, inline engine and not the rear-mounted, air-cooled, flat powerplant: The latter characteristics are a tradition at Porsche and even the vaunted 917-30 Can-Am racer relied on an air-cooled, opposed 12-cylinder engine for power. Like the 914, the racer's engine was mounted midship, not in the rear, but the point is, Porsches always had their powerplants behind the driver. The 911 and 930 still do.

So the iconoclast 924 was not at all likè its predecessors and, horror of horrors, it was barely a Porsche. The 4-cylinder sohc engine was an Audi unit, and the few chassis components that were not Audi parts turned out to be Volkswagen items. If that's not enough, the car was being built in the Audi plant at Neckarsulm, West Germany. Ah well, at least the seats looked like Porsche designs.

In spite of all this the 924, originally developed for Volkswagen but "bought back" by Porsche after VW management decided against producing it, caught the motoring public's fancy. Ours too. In fact, in a 3-car comparison test (R&T July 1976) the Porsche beat the Datsun 280Z and the Alfetta GT. Our testers lauded its seats ("good lateral support with proper lateral retention"), instrumentation ("offers more information, all easy to read"), handling ("a joy, well balanced and flat") and steering ("light and quick").

With such attributes one would expect the 924 to be a fine little GT. Yet it falls short of the mark because of a number of niggling problems. First, there is the car's ride, which is jouncy and harsh. Porsche has been working on this problem and has alleviated one aspect of it—the "freeway hop" that plagued the 1976 models.

5

The 924's cockpit is generally well laid out and nicely finished, but the small vertical clearance between steering wheel and driver's seat, for which Porsche tried to compensate with a not-quite-round wheel, causes knee-to-wheel interference.

The 2 + 2 rear seating is, for the American market, officially not seating; no seatbelts are included.

The factory tightened up installation tolerances and (for 1978) installed hydraulic transmission mounts and new rubber rear suspension mountings, making the ride level if not smooth. These got rid of the hop but did little to improve compliance (the ability to soak up small but sharp bumps), which is still rather poor. The manufacturer has also failed to remedy another suspension-related problem, the booming that occurs when the car runs on very coarse road surfaces: the car's stiff suspension transmits the jarring action of the wheels through the springs and torsion bars into the body.

The 924's other major shortcoming is its engine, which is rather flat and noisy. Again, it's something Porsche has been working on and will solve shortly with turbocharging. In mid-1977 the 1984-cc four gained 15 more horsepower after Porsche raised the compression ratio from 8.0:1 to 8.5:1, increased the size of the intake valves from 38 to 40 mm and jiggled camshaft and ignition timing. Around-town performance improved (raising the rear axle ratio numerically undoubtedly helped too), but in the quarter mile the 110-bhp 924 is only 0.3 seconds faster than the 1976 version with an elapsed time of 18.0 sec, not very quick for a car of its stature. The same buzzy engine noise that marred the earlier cars' image remains: The show begins at about 4000 rpm and continues on up toward the redline of 6500 rpm. We called it "noisy, noisy, noisy" in our test of the 1977½ and we'll have to say the same thing about the current car.

Yet, in spite of its shortcomings, a Porsche is still a Porsche, even if it is built in Neckarsulm and not Stuttgart. Quality control (Audi's) is good, in the Teutonic tradition. And the car's styling is still as fresh as it was at birth: a cleanly designed car inside and out. Its lines are distinctly Porsche, meaning that at a glance the uninitiated observer would be hard-pressed to determine if the engine is in front or in the rear. The headlights fold down into the body when not in use and the all-glass rear hatch opens to allow access to an expansive package shelf which now has a vinyl roll-up security shade.

The interior itself was best described by one staffer as "functional, austere, Teutonic." That says a lot but it doesn't explain that the 924's interior is designed with the driver in mind. Witness the well placed gauges and easily reached controls, all

CONTINUED ON PAGE 25

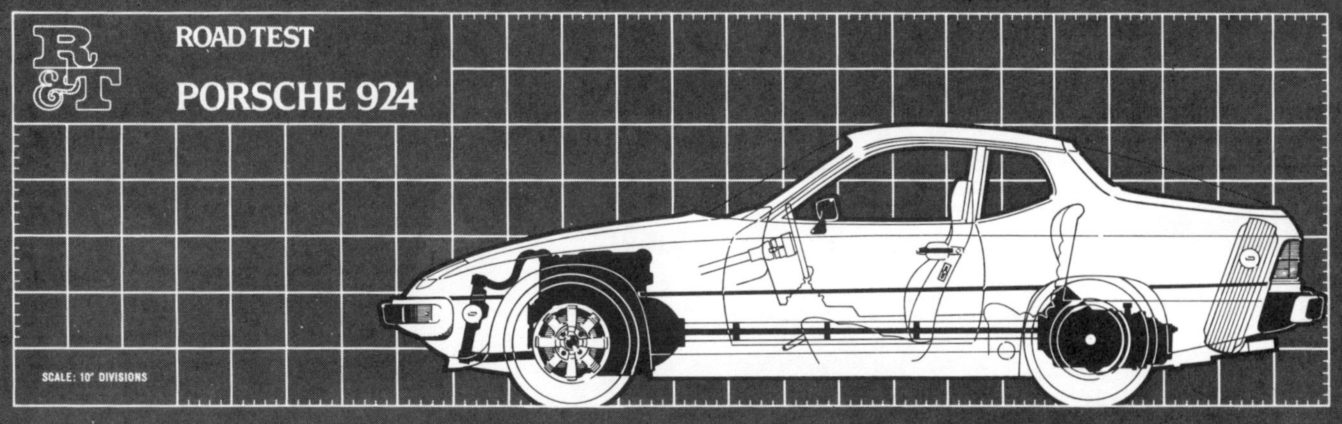

PRICE

List price, all POE $11,995
Price as tested $14,449
Price as tested includes: metallic paint ($395), air cond ($595), AM/FM stereo ($199), anti-roll bars ($150), Touring Package I ($520), removable sunroof ($380), tint glass ($105), Calif. emissions ($110)

IMPORTER

Porsche-Audi Div, VW of America
818 Sylvan Ave.
Englewood Cliffs, N.J. 07632

GENERAL

Curb weight, lb 2575
Test weight 2715
Weight distribution (with driver), front/rear, % 49/51
Wheelbase, in. 94.5
Track, front/rear 55.9/54.0
Length 170.1
Width 66.3
Height 50.0
Ground clearance 5.9
Overhang, front/rear 34.4/41.2
Usable trunk space, cu ft .... 10.4+7.9
Fuel capacity, U.S. gal. 16.4

ENGINE

Type .... sohc inline 4
Bore x stroke, mm .... 86.5 x 84.4
Equivalent in. .... 3.41 x 3.32
Displacement, cc/cu in. .. 1984/121
Compression ratio 8.5:1
Bhp @ rpm, net .... 110 @ 5750
Equivalent mph 99
Torque @ rpm, lb-ft .. 111 @ 3500
Equivalent mph 68
Fuel injection .... Bosch K-Jetronic
Fuel requirement ..unleaded, 91-oct
Exhaust-emission control equipment: catalytic converter, air injection, exhaust-gas recirculation

DRIVETRAIN

Transmission .... 4-sp manual
Gear ratios: 4th (0.97) 3.76:1
3rd (1.36) 5.28:1
2nd (2.13) 8.26:1
1st (3.60) 13.97:1
Final drive ratio 3.88:1

ACCOMMODATION

Seating capacity, persons ...... 2+2
Seat width, f/r, in....2 x 19.5/2 x 15.0
Head room, f/r 37.0/32.0
Seat back adjustment, deg 50

CHASSIS & BODY

Layout .... front engine/rear drive
Body/frame .... unit steel
Brake system .... 10.1-in. discs front, 9.1 x 1.5-in. drums rear; vacuum assisted
Swept area, sq in. 269
Wheels .... cast alloy, 14 x 6J
Tires .... Uniroyal Rallye 240, 185/70HR-14
Steering type .... rack & pinion
Overall ratio 19.2:1
Turns, lock-to-lock 4.0
Turning circle, ft 30.3
Front suspension: MacPherson struts, lower A-arms, coil springs, tube shocks, anti-roll bar
Rear suspension: semi-trailing arms, torsion bars, tube shocks, anti-roll bar

INSTRUMENTATION

Instruments: 150-mph speedo, 7000-rpm tach, 99,999.9 odo, oil press., coolant temp, voltmeter, fuel level, clock
Warning lights: oil press., brake system, handbrake, fog lights, exh-gas recirc, seatbelts, hazard, high beam, directionals

MAINTENANCE

Service intervals, mi:
Oil change 7500
Filter change 15,000
Tuneup 15,000
Warranty, mo/mi 12/20,000

CALCULATED DATA

Lb/bhp (test weight) 24.7
Mph/1000 rpm (4th gear) 18.9
Engine revs/mi (60 mph) 3180
Piston travel, ft/mi 1760
R&T steering index 1.23
Brake swept area, sq in./ton .. 198

RELIABILITY

From R&T Owner Surveys the average number of problem areas for all models surveyed is 12. An average of 7 of these problem areas is considered serious enough to constitute reliability areas that could keep the car off the road. As owners of earlier-model Porsches reported 11 problem areas and 4 reliability areas we expect the overall reliability of the Porsche 924 to be better than average.

ROAD TEST RESULTS

ACCELERATION

Time to distance, sec:
0-100 ft. 3.5
0-500 ft 9.7
0-1320 ft (¼ mi) 18.0
Speed at end of ¼ mi, mph 77.0
Time to speed, sec:
0-30 mph 3.3
0-40 mph 5.3
0-50 mph 7.7
0-60 mph 11.0
0-70 mph 14.7
0-80 mph 19.4
0-90 mph 27.2

SPEEDS IN GEARS

4th gear (6200 rpm) 117
3rd (6500) 86
2nd (6500) 55
1st (6500) 34

FUEL ECONOMY

Normal driving, mpg 18.5
Cruising range, mi (1-gal. res)....285

HANDLING

Speed on 100-ft radius, mph .. 33.9
Lateral acceleration, g 0.766
Speed thru 700-ft slalom, mph....58.9

BRAKES

Minimum stopping distances, ft:
From 60 mph 150
From 80 mph 280
Control in panic stop ....very good
Pedal effort for 0.5g stop, lb 25
Fade: percent increase in pedal effort to maintain 0.5g deceleration in 6 stops from 60 mph .... nil
Overall brake rating..........very good

INTERIOR NOISE

All noise readings in dBA:
Idle in neutral 56
Maximum, 1st gear 92
Constant 30 mph 70
50 mph 75
70 mph 80

SPEEDOMETER ERROR

30 mph indicated is actually 28.0
50 mph 47.0
60 mph 57.0
70 mph 66.5
80 mph 76.0
Odometer, 10.0 mi 9.8

ACCELERATION

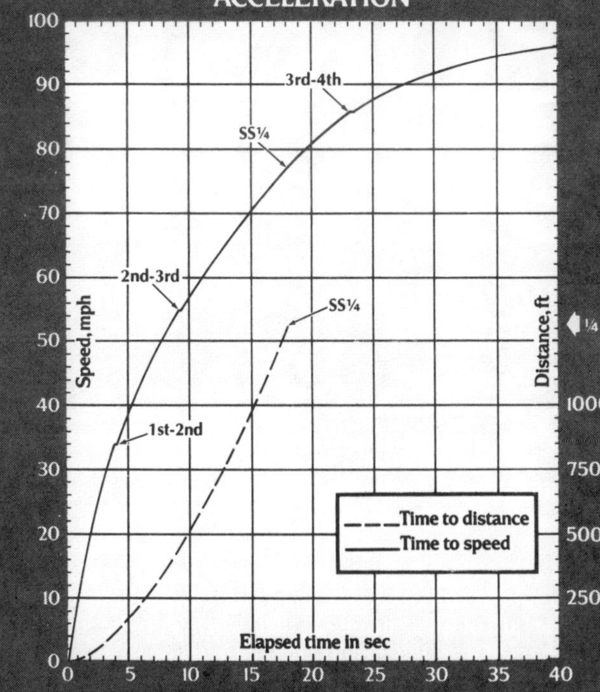

PORSCHE 924 TURBO

Zuffenhausen's sensational remedy for the blahs

BY JOE RUSZ

PHOTOS BY THE AUTHOR

PORSCHE HAS FINALLY found a cure for the 924's less than spectacular performance. The remedy is turbocharging, the automotive world's latest panacea. These days everyone is building turbocars, but the Zuffenhausen-based firm was the first to popularize the concept and make it really work. It's no wonder they made it look so easy. With years of experience in Can-Am and GT racing, Porsche probably has more cumulative knowledge about turbocharging than any other automobile builder. What all of this means is that when Porsche builds a turbocharged car the results are outstanding. We said so after driving the 930 Turbo Carrera. We're saying it again, having recently driven the 924 Turbo, the 4-cylinder baby brother of the Carrera.

When we told Porsche's engineers how impressed we were with their latest design, the expression on their faces said, "Did you expect anything less from us?" Frankly, yes. For more than three years we have eagerly awaited improvements in the 924's performance, which has often been eclipsed by cars costing much less. We've also complained about the car's ride and noise level, and though the factory made some changes to boost horsepower and dampen freeway hop (R&T, August 1977), we felt the car was still slow and noisy. Having driven the 924 Turbo, we can happily report that this is no longer the case. Not only does turbocharging make for a quieter running engine, it also breathes life into a powerplant that labors badly in its normally aspirated U.S. emissions trim. Dr Ernst Fuhrmann, Porsche's President and

Chairman of the Board, says that turbocharging and the other changes represent the first steps in the evolution of the 924. He calls this a modest improvement, but that's just an example of the factory's penchant for understatement.

On the surface, the 924 Turbo may indeed appear to be just a modest exercise in styling. There are no flared fenders (though these will undoubtedly appear in the future, at least on the racing versions), whale tails or other gimmicks. Porsche says it "intentionally avoided playing up the high-performance characteristics of this new (automobile) with brash styling tricks." How true. The first tipoff it's a Turbo are slots and louvers in the front spoiler and more louvers in the sheet metal just above the front bumper. The bottom openings direct air to the oil cooler and the brakes, while the top ones exhaust hot air from the engine compartment. A NACA duct in the hood also helps remove hot air from the area around the turbo which is located on the right side of the engine. At the rear of the car, a rubber spoiler has been permanently bonded to the glass hatch to give the car added stability at high speeds. Sorry, but the spoiler can't be retrofitted to earlier 924s. From the side, special cast alloy 6 x 15-in. spoked wheels tell you it's a Turbo. And lest you be fooled by imitations which undoubtedly will soon flood the market, the factory wheels have a 5-bolt pattern identical to the 911's. Other touches include an electrically operated side view mirror, but this isn't unique as it is an option on the present model. What is unique to the Turbo is an optional 2-tone paint scheme that Porsche says it's trying ⟫

out just to make the new model more distinctive, at least in the beginning. They may like it, but most of the American journalists attending the long-lead preview in Friedrichsruhe, West Germany, found the contrasting top and side colors garish. Oh, there's a 924 Turbo badge on the wraparound, but that's about it.

On the surface, the Turbo may look like nothing more than a customized 924, but don't be fooled. Beneath the sheet metal are significant changes in the engine, driveline, running gear and suspension. The 924 Turbo is a total performance package with all the requisites one expects to find in a Porsche.

Let's start with the engine. It's the familiar sohc inline 1984-cc four that has powered the car since its inception, but with these differences. There are now forged pistons and while the forged connecting rods and crankshaft and cast iron cylinder block are the same as before, the cylinder head isn't. The combustion chamber of the normally aspirated engine is totally within a recess in the piston crown; the new 924 head is a pent-roof design with the combustion chamber partially in the head and partially in the piston. The sparkplugs have been relocated from the exhaust to the intake side of the head, and this change results in a more favorable combustion cycle with lower exhaust gas temperatures and consequently better performance, according to 924 Turbo project engineer, Paul Hensler. Although the camshaft is identical to the one used in the normally aspirated engine, the diameter of the turbo engine's exhaust valves has been increased 3 mm. Breakerless electronic ignition replaces the point-type used on the conventional engine and the new unit has a built-in rev limiter that cuts off the spark at anything over 6500 rpm. Bosch platinum-tipped sparkplugs are used because Porsche feels only these plugs can survive the rigors of turbocharging. Incidentally, final assembly and dynamometer testing of the engine takes place at Porsche's factory in Zuffenhausen rather than Audi's facility in Neckarsulm.

The Turbo engine's 7.5:1 compression ratio is a point lower than the ratio of the normal 924 powerplant, but it's still relatively high for a turbocharged engine developing 10 psi of boost. Under full boost conditions the effective compression ratio jumps to 10.8:1.

Of course, the heart of the new powerplant is the KKK turbocharger located at the lower front corner of the engine. Porsche gave its placement a lot of thought because it wanted the unit close to both the exhaust header and the Bosch K-Jetronic's fuel metering device. To achieve the desired location, the alternator had to be relocated from the lower left to the upper right front of the engine block. The K-Jetronic's fuel meter was moved, the diameter of the fuel lines was increased and a second fuel pump located in the trunk was added. Such subtle but necessary design changes set the factory's Turbo apart from a typical aftermarket installation.

The factory uses a wastegate located upstream of the compressor to regulate boost because this is the most desirable method of doing so. It's also more expensive. That's why most bolt-on kits use a pop-off valve, located downstream of the turbo to control boost by restricting the flow of air entering the intake manifold.

Something else you're not likely to find on an aftermarket kit is a second protective device, a bypass valve that is integrated into the compressor housing. Porsche calls it a circulating valve and says it protects against sudden surges in pressure should the throttle close abruptly. When that happens, the pressurized air-fuel mixture is routed back through the compressor and continues circulating through this closed loop made up of the compressor and the valve.

When the turbocharger gets to spinning (at speeds up to 100,000 rpm) and the boost peaks, the European Turbo develops 170 bhp DIN at 5500 rpm. By comparison, the normally aspirated European engine produces 125 bhp at 5750 rpm. Torque is also increased and now measures 180 vs 122 lb-ft at an identical (for both engines) 3500 rpm.

This substantial increase in horsepower and torque calls for improvements in the driveline and transaxle, so Porsche has increased the driveshaft diameter from 20 to 25 mm and strengthened the rear halfshafts. The factory has also beefed up the various gears and shafts in the transaxle which is now a 5-speed design. The ratios are tailored to the turbo powerplant and are much higher (lower numerically) than the non-turbocharged car's. The rear axle ratio is also taller, 3.17:1 instead of 3.44:1. What this means is that in top gear, the 5-speed's overall ratio is 2.25:1 vs 3.34:1 for the 4-speed and the engine just loafs along. As expected, this tall gearing gives the European 924 Turbo an exceptionally high speed—more than 150 mph! But before you get too enthusiastic, Porsche also says the U.S. version will get shorter final drive and gear ratios and thus a lower top speed that will coincide with "American driving habits."

Up to this point we have a car that accelerates well and has a remarkable ability to reach high terminal speeds. Isn't that all we've asked for in the 924? Perhaps, but to cope with the higher level of performance, the factory fitted 4-wheel-disc brakes to the 924 running gear. The brakes and hubs are 911 units which explains why the new wheels have a 5-bolt pattern. Are 911 wheels adaptable to the 924 Turbo? Not at this time, because the

<hr>

PORSCHE 924 TURBO—EUROPEAN SPECIFICATIONS

GENERAL

Curb weight, lb/kg	2600	1180
Wheelbase, in./mm	94.5	2400
Track, front/rear	55.8/54.8	1418/1392
Length	165.8	4212
Width	66.3	1685
Height	50.0	1270
Fuel capacity, U.S. gal./liters	16.6	63

CHASSIS & BODY

Layout	front engine/rear drive
Brake system	vented discs front & rear

Wheels	cast alloy, 15 x 6J; optional 16 x 6J
Tires	185/70VR-15; optional 205/55VR-16
Steering type	rack & pinion
Turns, lock-to-lock	4.0
Suspension, front/rear:	MacPherson struts, lower A-arms, coil springs, tube shocks, anti-roll bar/semi-trailing arms, torsion bars, tube shocks, anti-roll bar

ENGINE

Type		sohc inline 4
Bore x stroke, in./mm	3.41 x 3.32	86.5 x 84.4
Displacement, cu in./cc	121	1984
Compression ratio		7.5:1
Bhp @ rpm, DIN/kW		170/125 @ 5500
Torque @ rpm, lb-ft/Nm		180/245 @ 3500
Fuel injection		Bosch K-Jetronic

DRIVETRAIN

Transmission	5-sp manual
Gear ratios: 5th (0.71)	2.25:1
4th (0.93)	2.95:1
3rd (1.22)	3.87:1
2nd (1.78)	5.64:1
1st (3.17)	10.05:1
Final drive ratio	3.17:1

offset is different. But there are optional cast alloy 16 x 6J wheels fitted with Pirelli P6 205/55VR-16 tires. Heavy-duty shock absorbers are used as are front and rear anti-roll bars with diameters of 23/14 mm vs 20/18 mm. Porsche says other changes such as spring rates are minor and are only designed to tune the chassis to its new powerplant.

The interior is essentially standard 924 fare, but there are those little touches that set the Turbo apart from other 924s. The most obvious one is the steering wheel, the 3-spoke Carrera-type borrowed from the 911SC. It has much less offset than the old 924 wheel which tended to clash with one's thighs when it was turned 180 degrees. Besides the new 3-spoke wheel, Porsche is also offering an optional 4-spoke design. Undoubtedly, one or both of these wheels will be offered on the non-turbocharged 924.

Other standard interior touches include leatherette upholstery with cloth, herringbone tweed accents; electric window lifts; and a console with a voltmeter, clock and oil pressure gauge. The dash instruments include a 260-km/h speedometer (the U.S. version will probably get a 150-mph speedo) and a 7000-rpm tachometer with a 6500-rpm redline. The tach has been re-designed so that the redline is now at 12 o'clock. There is no boost pressure gauge because Porsche claims nobody ever looks at it anyway. We disagree—enthusiast drivers do look at their boost gauge and even use it to monitor engine conditions. The 5-speed's shift pattern is something else that many of us at R&T take issue over. The shift lever follows the 928's pattern with 1st gear down and to the left, reverse above it and 5th down and to the right. Most of us prefer 5th out of the way—up and to the right, as it is on the 911 and other 5-speeds.

Driving Impressions

SUBTLE IS not the word one would use to describe the Turbo 924's performance. Sensational is more like it. When the engine is started, the first thing you notice is its quietness. Porsche says, "The turbine absorbs a portion of the thermic and dynamic energy from the exhaust gas driving it and thus has the effect of an additional muffler." True, but more than that, the increased horsepower and torque make the car easier to drive. The engine produces as much bhp and torque at 1500 rpm as the normally aspirated powerplant, so there's no loss of low-end flexibility as there is with most other turbos. Rolling away in 1st gear, everything seems normal until about 3500 rpm when whoomp, the turbo comes on boost. Then, as with Porsche's other Turbo, there's that characertistic dizzying climb up to the redline. A quick snick into 2nd, then 3rd gear produces more of the same, but by the time you reach the top two gears and speeds well over 100 mph, the effect becomes more subdued. At high cruising speeds, it is almost impossible to discern the presence of a turbocharger. Occasionally, there's a faint whine from the engine compartment, but only the attuned ear will detect this not unobjectionable note. Engine noise is vastly reduced too.

Our European test car had no emission controls so the unfettered engine revved freely, right up to the rev limiter. Driveability was very good and the transition from no boost to boost was much less noticeable than with the 930 Turbo Carrera. Incidentally, Porsche indicates that driveability of the U.S. version, which will use a smaller turbo and possibly an even higher compression ratio, will be minimally affected. The factory has wisely chosen to use an oxygen-sensing system with a 3-way catalyst to control emissions. How much a reduction to about 150 SAE net bhp will have on the performance of the U.S. Turbo is hard to estimate because (in addition to other things) the factory hasn't finalized the transmission and final drive ratios.

With its increased horsepower the true handling potential of the 924 becomes apparent. Although mild understeer is the car's normal attitude, controllable oversteer can be induced as more power is applied.

Keeping the 924 Turbo under control at high speeds was no problem. The handling is excellent. It's stable and predictable. On one occasion, I lifted off the throttle in a high speed turn as a slower car pulled in front of me. This sudden decrease in power might have caused a frantic moment in other cars, but in the 924 Turbo, there was no change in attitude. In short, the new Porsche is a very reassuring car to drive at extremely high speeds.

The ride is very firm but with reasonable compliance; the sport shocks are effective, even if they are a bit stiff. What's impressive though, is the overall tightness of the chassis/body, even over rough surfaces. No rattles, squeaks, not even that dreadful booming that has plagued the 924 since its introduction.

After completing the first half of my evaluation drive over twisty, 2-lane country roads, I headed for the *Autobahn* for the acceleration and top speed runs. Both were simply incredible. With two people aboard, the Turbo needed just 7.6 sec to go from 0 to an indicated 100 km/h (62 mph). Then in 5th gear, it was an astounding experience to watch the tach needle climb slowly toward the redline as the speed climbed—220 km/h, 240 and finally an indicated 250! The engine was still pulling but slower traffic forced me to slow down. It was later at the press conference that I learned the car's real top speed is 243 km/h—more than 150 mph. Whew! Interestingly, the factory gives 225 km/h as the top speed of the European Turbo. Why so low? "Because if we quoted a higher number and then had a customer discover his car could not attain that speed, we would have a very unhappy owner," explained Dr Fuhrmann.

Getting a turbocharged car up to 150 mph is one thing, keeping it there, on boost for sustained periods, is another. With the 924 Turbo this was never a problem. During my blast down the *Autobahn* the engine never faltered and the coolant temperature never reached the halfway mark on the gauge, even though the car's speed didn't drop below 120 mph for perhaps 10 miles. I asked Porsche about such driving and was told that a test car driven from France to Germany at never less than 120 mph, showed no signs of detonation or any of the other ills associated with prolonged turbo use. For U.S. buyers though, the question is, how will the use of unleaded, low-octane fuels affect this reliability?

The only distressing news about the 924 Turbo came when I asked the price. Like the car's top speed, it too is high—39,480 Deutsche Marks or about $21,340 for the European model, which will appear in February. The U.S. version which will go on sale in August or September (probably as a 1980 model), will cost between $19,000 and $20,000. I asked Dr Fuhrmann about the high price of all Porsches and he explained that Porsche's goal is to build fine automobiles without compromises in quality and performance. In today's market, this cannot be done inexpensively, especially in Germany.

Porsche plans to build about 20 Turbo 924s per day (20 percent of that model's production) at the Audi plant in Neckarsulm. At the same time they will continue building 911SCs, Turbo Carreras and 928s in Zuffenhausen. But ultimately, the 911SC will be phased out—Fuhrmann says, only when it can no longer meet emissions and safety specifications—and the 924 will become the bread-and-butter model of the Porsche line. There will be a variety of models, some of which are already in the works. For instance, there will soon be four different U.S. 924s—a normal car with and without a turbo, a car with the improved chassis but a non-turbocharged engine and a full-fledged Turbo. By the way, Porsche says the improved 924 without turbo should appeal to SCCA D Production racers who have been complaining of brake fade with the present front disc/rear drum brakes.

Manfred Jantke, Director of Public Relations and Racing for Porsche, elaborated on the future of the 924 in racing. He said that in time the competition 924 will become the sort of "affordable" racer that the 911 Carrera RSR once was. This car cost about $30,000 back in 1973, a marked contrast to the 935 Turbo which costs approximately $120,000. Jantke says the full-blown, large-displacement turbos like the 935s will be discontinued in 1981 after new European production car racing rules take effect.

As for the 924 road cars, they will continue to improve as the factory puts its weight behind the 924 design. The firm might consider the 924 Turbo just a "modest change," but it's enough to make us exclaim, "At last, here's a 924 that deserves to be called a Porsche."

PERFORMANCE COMPARISON
Production & Supercharged Porsche 924s

	Production	Supercharged
Acceleration:		
Time to distance, sec:		
0–1320 ft (¼ mi)	18.0	16.0
Speed at end of ¼ mi, mph	77.0	88.0
Time to speed, sec:		
0–30 mph	3.3	2.4
0–60 mph	11.0	7.7
0–80 mph	19.4	12.8
0–90 mph	27.2	16.8
Fuel economy:		
mpg	18.5	11.5
Interior Noise:		
Idle in neutral, dBA	56	70
Constant 50 mph	75	82
70 mph	80	81

BUTERA'S SUPERCHARGED 924

Blown out of proPorscheon

BY GLENN BRINKS
PHOTOS BY JOE RUSZ

JOHN BUTERA'S PORSCHE 924 is supercharged. That's right, supercharged, not turbocharged. Supercharging is usually associated with drag racing rather than sports cars, and that's where Butera comes into the picture. Until he left drag racing three years ago, he was known as the premier funny car builder in the country. Butera's funny cars were raced by some of the best drag racers—Tom McEwen and Don Prudhomme among others. Butera left drag racing to do consulting for Revell Models, build street rods (perhaps as much admired as his race cars) and work on personal projects. The 924 idea started because, "I saw one. I liked the looks, so I bought it." Not satisfied with the 924's 110 bhp, he added a supercharger because he knew that on a dragster or street rod a supercharger increases horsepower, gives a wide powerband with plenty of low-rpm torque *and* provides fast throttle response.

In the past, those same qualities made superchargers extremely successful in road racing and sports cars. Some are well known, such as the Mercedes 540K and the Blower Bentley. Less well known is that Fiat was winning Grands Prix with its supercharged Type 805 as early as 1923, and from then until World War II, nearly every major Grand Prix was won by a supercharged car.

After the war, road racing rules were changed, favoring normally aspirated engines, and supercharging lost popularity. There were still a few supercharger kits available, such as the Judson kit R&T tested on an MGTF in November 1954, and there were a few specials such as Carroll Shelby's personal 427 Cobra with twin Paxton superchargers (R&T, February 1968), but in the Seventies turbochargers took over and engine-driven superchargers faded from view. It remained for a drag racer to revive the tradition of supercharged sports cars.

Butera is known as one of the best "combination" men (designer, fabricator, welder, machinist) in the business and his Porsche shows it. Under the hood there's no clutter and everything added shows an aircraft-like finish of brushed aluminum with surgically neat welds. Dominating the scene is a Keith Magnuson-built Magnacharger Roots-type supercharger originally developed for drag racing motorcycles. It's supported by a custom-fabricated aluminum manifold and the drive is supplied by a toothed belt and special pulleys. Bolted to the other side of the supercharger is an unexpected item on a Porsche 924—a Weber 40 DCMF 12 carburetor. Butera, experienced in carburetor tuning, felt that development would be easier with the carburetor than with a fuel-injection system that was unfamiliar to him. (A supercharged 924 using stock fuel injection will be his next project.) The other major addition is a water-injection system to prevent detonation by cooling the intake charge with a fine spray of water. Otherwise, the engine is totally stock. Even the emission control systems (except the fuel injection) are in working order.

Performance, however, is far from stock. The supercharged Porsche hustled from 0–60 mph in 7.7 seconds, 3.3 sec faster than stock, while 0–90 mph took only 16.8 sec compared to 27.2 for a production 924. On the dragstrip, the blown 924 did the quarter mile in 16.0 sec at 88.0 mph, while a stock Porsche takes 18.0 sec and reaches 77.0 mph.

But numbers alone, however impressive, don't describe the complete transformation of character the supercharger causes on

the 924. Punch the throttle at almost any speed and the car accelerates briskly. There's no waiting for the boost to build up as with a turbo. Acceleration is slower at low rpm, of course, but because the supercharger is driven directly by the engine, some boost is available throughout the rpm range, with about 7.0 psi developed at 6500 rpm. The subjective feeling is one of a steady force pushing you back into the seat, rather than the sudden kick in the back you get when a typical turbocharged engine comes on the boost.

Around town the broad powerband results in a car with few vices. Occasionally when moving away from a stop the engine would hesitate slightly but this can probably be cured with a little fine-tuning of the idle mixture. The supercharged engine makes for relaxed around town driving and makes merging with freeway traffic a breeze.

Out on a country road, the sheer eagerness of the engine can bring even the most sedate driver to life. Hurry up to a turn, touch the brakes, drop down a gear and the 924 simply rushes away, impatient for the next one. As two of the test drivers put it, "It runs like a well tuned small V-8," and "This is the way the 924 should have performed in the first place."

The price of this performance is lower fuel economy: Under moderately hard driving the supercharged 924 delivered 11.5 mpg. Gentle driving improves this, but it's still far below the 18.5 mpg recorded by the stock 924. Part of the reason for the lower fuel mileage is the supercharger: because it's mechanically connected to the engine it absorbs some power. Turbochargers, or more properly turbo-superchargers, typically have a smaller effect on fuel economy because the power loss caused by increased exhaust back pressure is usually far smaller than the power required to drive the compressor mechanically.

The water injection system used water rapidly during our tests, emptying its 1.5-quart container in as little as 34 miles, but it was improperly jetted. Butera says with proper jetting water consumption should be about 1.5 quarts per tankful of gas.

Along with the horsepower, the supercharger also adds a unique sound. It's a high-pitched mechanical whine remarkably like the sound of an ERA, a supercharged Mercedes GP racer or a jet plane coming in to land. In August 1977, we described the 924 as "noisy, noisy, noisy" and this 924 was worse, but not because of the supercharger. The exhaust system resonator was cut off for clearance during the installation of European-style bumpers front and rear, resulting in an irritating hole-in-the muffler sound that got on everyone's nerves and masked the far more pleasant sounds from the supercharger. And in one respect the supercharger helps reduce engine noise. Because the engine revs more quickly it spends less time buzzing through some of the irritating resonance bands that plague the stock engine.

In other respects, this 924 lives up to its engine. It has been lowered 2.0 in., and even more than the stock 924 Porsche, this one "sticks to the road like chewing gum on the bottom of a theater seat." We called the stock 924 seat "as comfortable as a Gucci loafer" but Butera went even further and installed a pair of Recaro buckets. The eccentric Porsche steering wheel was replaced with a leather-wrapped Personal model and even the instrument panel was modified. All removable dash panels were replaced with panels machined from aluminum, a trademark of a Butera-built car. Most staffers liked the functional styling but one thought a black anodized finish would show less dirt and glare. Complementing the stock instrumentation is a complete set of VDO instruments including fuel pressure and intake manifold pressure/vacuum gauges.

It's an impressive piece of work. Along the way, Butera had to solve problems such as uneven fuel distribution among the cylinders, overheating and short sparkplug life and he did it without internal engine modifications. Now that the development work has been done, any home mechanic should be able to bolt the pieces onto his own 924 in a few hours. For 924 owners who value performance over fuel economy, it's a serious alternative to a turbocharger kit. The complete supercharger kit sells for about $1400, with installation available for $200 more. Information kits are $2.00 and are available from John Butera, 3111 Walnut Ave, Huntington Beach, Calif. 92638; 714 960-4492. ⊗

Do they lose something in the transmission?

AUTOMOTIVE PURISTS MAY disagree, but there's a lot to be said for sports and GT cars equipped with automatic transmissions. Such cars are easier to drive (no fancy footwork or sleight-of-hand needed), smoother (a good torque-converter-equipped automatic makes almost indiscernible shifts) and less fatiguing than their standard-shift counterparts. The price of these cars puts them into another market class composed of people who often want the luxury and convenience of a Cadillac in something with more sporting pretensions. And with the lady of the house doing much of the driving, comfort and convenience take on added importance. So while we'll concede that the automatic sports car is not the ultimate go-fast road car, it is a valid compromise for the person who wants the looks (status), handling and performance of a sports/GT without the inconvenience of shifting.

In choosing the four candidates for this test, the Chevrolet Corvette, Datsun 280ZX, Mazda RX-7 and Porsche 924, we tried to arrive at a middle ground of performance, luxury and price. Meaning that cars like the Porsche 928 (base price $31,835) and the Triumph TR7 (base price $6995) were excluded because they fell beyond the median, especially in cost. Incidentally, the Triumph TR8 would have been a natural, but it's unavailable at this time. How does the Mazda manage to sneak into our comparison? Simple. It has the dubious honor of being a $7995 automobile that dealers are demanding and unfortunately getting a lot more for. In today's inflationary economy, the as-equipped *suggested retail* prices of our four contenders ranged from a low of $9455 for the Mazda to a high of $17,790 for the Porsche. The Corvette and the Datsun were within two hundred dollars of each other at $12,505 and $12,682, respectively. Yes, it is possible to buy all of these cars for less, but when looking for luxury, one doesn't skimp on details. As a result, three of the cars had sunroofs, two had cruise control and all were equipped with air conditioning.

While all four cars might be considered similar in concept (sleek, sporty, etc), they are dissimilar mechanically and even structurally. For instance, the Corvette is powered by a 5735-cc ohv pushrod V-8 mounted in a steel chassis and clothed by a fiberglass body, the Datsun by a 2753-cc sohc inline six installed in a steel unit body. The Mazda and the Porsche are unit body designs just like the ZX, but while the 924 uses a conventional, 1984-cc sohc four, the RX-

7 relies on the revolutionary rotary for power. All four cars have torque-converter-equipped, 3-speed automatic transmissions, but the Porsche's gearbox is mounted at the rear and is actually a transaxle.

Suspension design also varies with each car. While the three imports all have MacPherson strut front ends, each has a different rear. The ZX and the 924 use semi-trailing arms and tube shocks, but the Datsun has coil springs, the Porsche, torsion bars. And the Mazda because of its live rear axle has an unusual setup—lower trailing links, upper angled links, a

Watt linkage, coil springs and tube shocks. The sole U.S. entry has unequal length A-arms, coil springs and tube shocks up front and lower lateral arms, axle shafts acting as upper lateral arms, a transverse leaf spring and tube shocks at the rear.

Armed with this information it's tempting to pick the best car of the lot. We'll spare you the suspense and also let you know if you made the correct choice by telling you which car won—after we explain how the entries in our comparison tests are scored.

Each of the six test drivers was given a

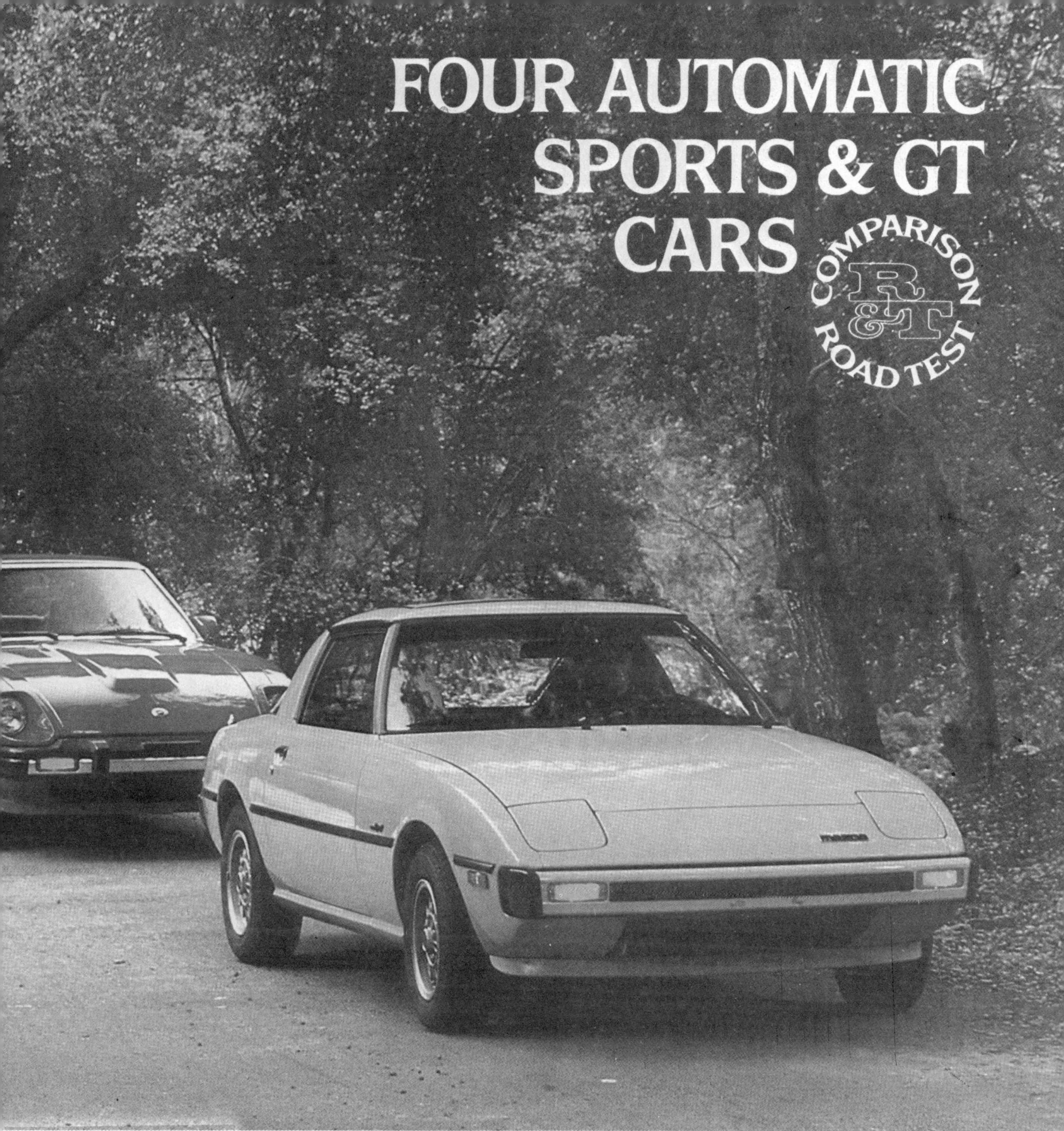

score sheet with 21 different categories including such obvious ones as handling, braking and ride and more esoteric qualities like interior styling, ingress/egress and body structure, to name a few. Each category was scored on a 1 to 10 basis with the highest number representing the best score. In addition to this numerical evaluation, each tester was asked to rate the four cars on a purely subjective basis, disregarding the computations. Thus it came as no surprise to anyone familiar with our comparison tests, that the highest scoring entry was not the most popular car. The Datsun won the numbers game accumulating 1054 out of a possible 1260 points. But the Mazda, a car almost not included in this test, won the hearts of all but one judge who, in spite of his preference for the ZX, readily admitted that considering its price, the RX-7 was undoubtedly the hands-down champ.

Statistically, here's how the cars fared. The Datsun was rated 1st by four of the testers (one tester had it tied with the Mazda), and 2nd and 3rd by the other two. The Mazda rated three 1sts, two 2nds and one 3rd. The Porsche had five 3rd places and one 2nd, while the Cor-vette got six 4th-place finishes. We'll tell you why after we outline our route.

Our 1020-mile test trek took us east from Dana Point, California to Blythe, California, then east-northeast to Prescott, Arizona. From Prescott, the course meandered west a bit before heading northwest to Kingman, Arizona. From Kingman, it jutted south a piece before turning west toward its Dana Point origin. Editor-at-Large, Henry N. Manney III, the Amerigo Vespucci of the automotive set, plotted the excursion which he himself did not attend. Undoubtedly, he knew what pitfalls would befall us—the ⟩⟩⟩

15

GENERAL DATA

	Chevrolet Corvette	Datsun 280ZX	Mazda RX-7 GS	Porsche 924
Basic price	$10,220	$9899	$7995	$14,600
Price as tested[1]	$12,505	$12,682	$9455	$17,790
Layout	front engine/ rear drive	front engine/ rear drive	front engine/ rear drive	front engine/ rear drive
Curb weight, lb	3655	2900	2435	2825
Test weight	3745	2970	2555	2925
Weight distribution (with driver), f/r, %	48/52	52/48	53/47	48/52
Wheelbase, in.	98.0	91.3	95.3	94.5
Track, f/r	58.7/59.5	54.5/54.3	55.9/55.1	55.9/54.0
Length	185.2	174.0	168.7	170.1
Width	69.0	66.5	65.9	66.3
Height	48.0	51.0	49.6	50.0
Usable trunk space, cu ft	10.4	19.8	15.0	10.4 + 7.9
Fuel capacity, U.S. gal	24.0	21.1	14.5	16.4
Brake system	11.8-in. vented discs front and rear; vacuum assisted	9.9-in. vented discs front, 10.6-in. discs rear; vacuum assisted	8.9-in. vented discs front, 7.9 x 1.3-in. finned drums rear; vacuum assisted	10.1-in. discs front, 9.1 x 1.5-in. drums rear; vacuum assisted
Wheels	cast alloy, 15 x 8	cast alloy, 14 x 6JJ	cast alloy, 13 x 5½J	cast alloy, 14 x 6J
Tires	Goodyear Polysteel, P225/ 70R-15	Toyo 702, 195/70HR-14	Bridgestone RD-106, 185/ 70HR-13	Pirelli CN36, 185/70HR- 14
Suspension, f/r	ind coil/ ind leaf	ind coil/ ind coil	ind coil/ live coil	ind coil/ ind tor

[1]As-tested price includes: for the Corvette, removable glass roof panels ($365), air cond ($635), tilt/telescoping steering wheel ($190), AM/FM stereo/8-track ($228), misc options ($784), Calif. emissions ($83); for the 280ZX, automatic transmission ($295), GL package including air cond, interior trim, pwr str, velour seats, AM/FM stereo, central warning system, cruise control ($2284), 2-tone metallic paint ($99), Calif. emissions ($105); for the RX-7, automatic transmission ($355), air cond ($555), aluminum wheels ($275), sunroof ($275); for the 924, air cond ($695), metallic paint ($395), leather seats ($680), sunroof ($395), power windows ($300), misc options ($615), Calif. emissions ($110).

ENGINE & DRIVETRAIN

	Chevrolet Corvette	Datsun 280ZX	Mazda RX-7 GS	Porsche 924
Engine type	ohv V-8	sohc inline 6	2-rotor Wankel	sohc inline 6
Bore x stroke, mm	101.6 x 88.4	86.0 x 79.0	240.0 x 70.0[2]	86.5 x 84.4
Displacement, cc	5735	2753	1146	1984
Compression ratio	8.9:1	8.3:1	9.4:1	8.5:1
Bhp @ rpm, SAE net	220 @ 5200	132 @ 5200	100 @ 6000	110 @ 5750
Torque @ rpm, lb-ft	260 @ 3600	144 @ 4000	105 @ 4000	111 @ 3500
Carburetion/Fuel injection	one Rochester (4V)	Bosch L-Jetronic	one Nikki (4V)	Bosch K-Jetronic
Fuel requirement	unleaded, 91-oct	unleaded, 91-oct	unleaded, 91-oct	unleaded, 91-oct
Transmission	3-sp auto	3-sp auto	3-sp auto	3-sp auto
Final drive ratio	3.55	3.55	3.91	3.73
Engine speed @ 60 mph, rpm	2700	2950	3700	3150

[2]Bore width x chamber major axis.

worst food this side of the military. Henry included plenty of what he calls "Targa Florio stuff," plus a bit of the Baja as well. Interstate travel and city driving were kept to a minimum as we all get plenty of that around the office. Elevations ranged from slightly below sea level to about 6000 ft above, not quite as high as previous trips which often reached 9000 ft. Weather conditions were favorable and while rain seemed imminent at times, it never materialized. By the way, former Editor Jim Crow, originator of the R&T comparison test, accompanied us on this trip and wrote the sidebar following this story.

Because there were plenty of desolate, flat stretches, each driver was able to evaluate things like high-speed stability, wind noise and top speed on a first-hand basis. At one point we encountered a strong cross wind which enabled us to check for wind wander. Interestingly, all four cars topped out at about 115 mph (indicated) even though their engine displacements ranged from 1146 to 5735 cc and horsepower ratings from 100 to 220 bhp. Although all four test cars registered about the same top speed, they were not always equal in other respects. Here are some of the specifics that made each entry a winner—or loser.

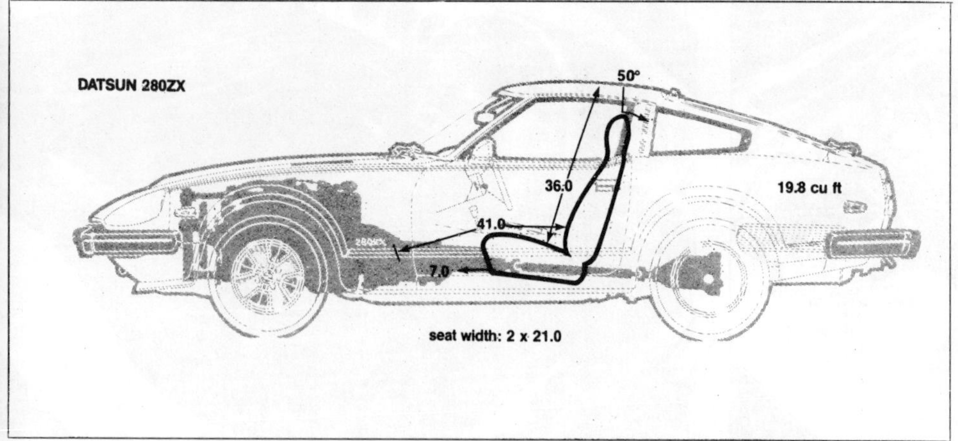

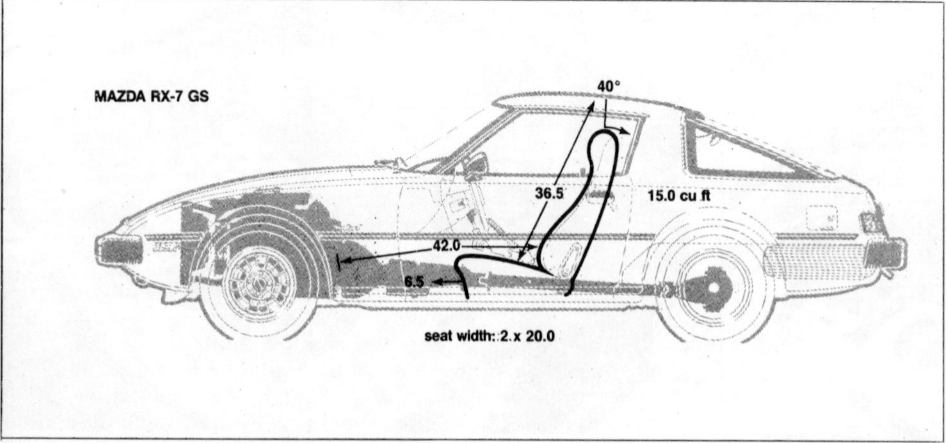

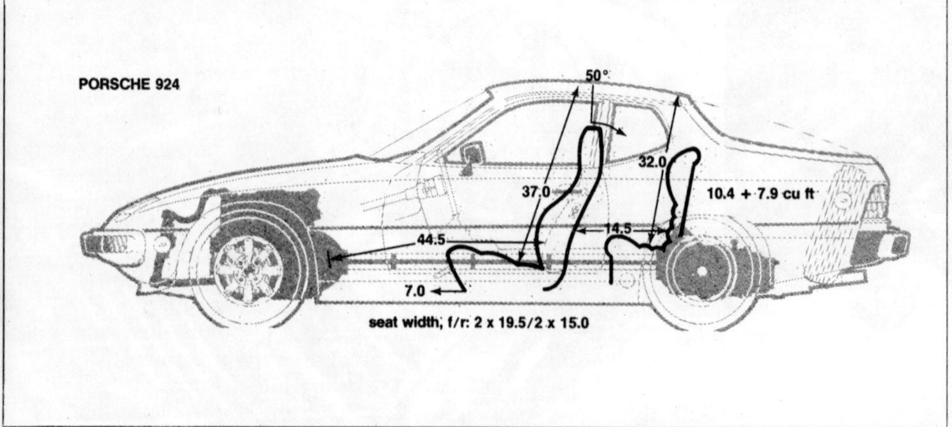

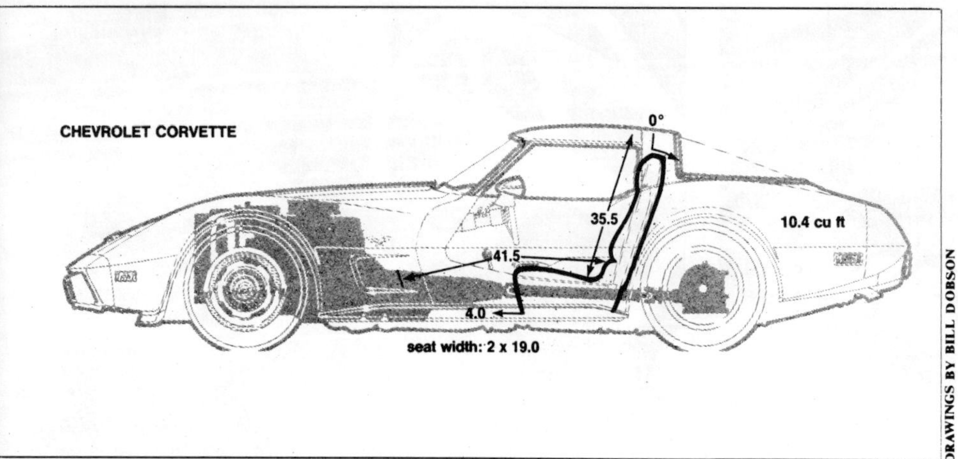

Datsun 280ZX

"THIS CAR continues to grow on me, which may be a reflection of how well it is designed," wrote one driver. Perhaps his statement explains how a car that was not the most popular choice scored the most points. The ZX was first in 10 of the 21 categories: ride quietness, controls, instrumentation, interior styling (where it tied the 924), interior finish, ingress/egress, luggage space and loading, body structure and seating comfort. Only two of these categories, ride and body structure, fall into the performance grouping, while the rest are part of either the comfort/controls/convenience group or the design/styling heading. The Datsun was not a stellar performer in the traditional sense (speed, cornering), but won everyone over with its amenities, layout, looks and comfort. "Luxury, opulence even," one tester wrote. "Quiet and comfy," said another. His statement was borne out by our figures. The ZX was first in the body structure category because it is a tight car. "Few rattles or creaks, solid on even the worst surfaces," was the way one of our staffers put it.

Our group considered the Datsun quiet "except for the wind rushing past the A-pillars at high speeds." Part of this results from the car's extremely clean lines, but much of it may be that the ZX was the only car without a sunroof, an option that produced at least a fair amount of wind noise and/or rattles on the other three cars.

The Datsun's controls and instrumentation scored highly, with good reason. The former, which includes such things as the headlight and windshield wiper stalks plus the heating/cooling levers and the shifter, are easy to reach. The latter, namely the speedometer, tachometer and engine gauges, are well placed and easy to read. And even the most jaded test driver didn't fail to be amazed or amused by the ZX's unique fuel gauge. Hand in hand with the controls and instrumentation are the Datsun's interior styling and finish, two more categories that were winners for the Nissan Motors entry. In fact the highest numerical score in any single category went to the Datsun whose interior finish earned 56 points.

Ingress/egress and seating comfort gave the Datsun two more wins as did luggage space and loading. The ZX rear cargo area is expansive (19.8 cu ft) and readily accessible through the rear hatch.

That's how the Datsun won the numbers game. Now here's why it wasn't the most popular car subjectively: It fell short of the mark in the handling, braking, steering, gearbox and engine categories, the traditional areas in which most of us judge such cars. The ZX was 3rd in all of the above, but was least criticized for its handling. Although one driver said it was sloppy at high speeds, others found the ZX quite stable, even at 90-plus mph.

The problem, it turns out, was not in the handling but in the power-assisted steering. "Light and vague," and "numb" were two of the descriptives used by our drivers. Most of us didn't like the car's unusual steering wheel which has its spokes at seven and five o'clock. As a result, the spirited driver has no place to hook his thumbs if he intends to assume the proper driving position. In time, most of us grew accustomed to the ZX's handling. "It's a car that can be hustled along quite nicely, once you get used to the steering."

Our drivers criticized the Datsun for its braking (light, with traces of fade after hard use), but complained most vehemently about the car's engine. "The engine runs out of breath at high rpm," was a complaint echoed by most of us. And we found that after a brisk workout, it would sputter and spit back through the injection on restart.

Yet, the Datsun scored the most points in spite of a poor showing in the above categories. Here's how one of our testers explained it: "The ZX pampers the driver with comfort and refined surroundings." How true and how appropriate for a car trading off the Gran Turismo image.

Mazda RX-7 GS

ONE DRIVER called it "The all-around athlete of the group," with good reason. The RX-7 was a spirited performer whose twin-rotor Wankel engine exhibited good flexibility and revved freely to its redline. "A torquey mother," said one of our testers. Another was more explicit: "It's smooth and plenty quick." From a standing start the RX-7 accelerates smoothly to its 7000-rpm redline, ticking off 0 to 60 mph in 8.9 sec and the quarter mile in 17.9 sec. Yet it has the ability to loaf along in high gear without stumbling. It's easy and fun to drive fast. Even so, the Mazda was not the quickest car of the group. That distinction went to the Corvette and while it's true the Vette was the champ in the 440-yard dash, performance is much more than just wind sprints.

The Mazda scored only three 1sts: steering, engine and driving position. Yet it accumulated 1033 points (21 less than the Datsun) by virtue of its dozen 2nd-place finishes and six 3rds. It fared extremely well in the performance areas which also explains why it was the most popular car on a purely subjective basis. "This is it," said one driver. "Easily my top choice," said another. Here's why they said what they did:

The RX-7's steering is one of its strong points. It's precise and accurate, with excellent feel. There is sufficient feedback to give you a good feel of the road, yet the steering effort is light enough so that the car gets along nicely without power assist. Although one test driver said the steering wheel's surface was too hard and too slick, he and the others liked the wheel's design and size.

A single point kept the Mazda from scoring a 1st in handling. One driver gave the RX-7's handling a solid vote of confidence, the rest rated it equal to or slightly below the Porsche's. But numbers aside, most of us were pleased with the Mazda's road manners. The car has mild understeer, but it's not enough to make spirited driving uncomfortable. All of us agreed that the RX-7 is surefooted and predictable and complimented Toyo Kogyo for doing it all with a live axle. "Crisp," was the way one of the drivers summarized the car's handling.

The RX-7's driving position scored three 1sts and three 2nds, giving the car a solid 1st in this category. This rather nebulous area is a bit difficult to define, since it encompasses other things such as seat comfort and controls, two categories in which the Mazda finished 2nd. If we have to pin it down, let's just say that driving position refers to such things as how the arms and legs reach the controls, how the body is supported in the seat, and how the driver is located relative to the steering wheel, windshield and floor. To paraphrase a popular beer commercial: "When it's right, you'll know it."

Other areas where the Mazda scored highly are listed in the chart, so we'll mention just the noteworthy ones. These include a roomy interior with a fairly large rear cargo deck (accessible through a rear hatch that can be opened from inside), well placed controls (stalks for the lights, wipers, etc), and a good exterior finish. But not necessarily its styling. "Too bad it's not better looking," was the way one driver put it.

Complaints? One major one, the RX-7's ventilation system. On an air conditioned car there is no way to direct air into the cabin without turning on the fan which also switches on the compressor. Turning on the compressor puts a load on the engine causing a decrease in performance and an increase in fuel consumption and this in a car with a small fuel tank.

The well intentioned folks at Toyo Kogyo equipped our test car with a removable, non-electric sunroof. Little did they realize that the device would prove to be one of the car's few annoying features. When the door was slammed (with the other door and the windows closed), air pressure caused the roof to pop up slightly. On rough roads the roof panel and the rear hatch creaked, probably as a result of a reduction in the car's torsional rigidity. This should explain how an otherwise quiet automobile failed to score a 1st in quietness.

In spite of these niggling problems, the Mazda still won the acclaim of almost all of us. "It's so good you have to look for things to pick on," one wrote. Another put it succinctly: "It's not as luxurious and well appointed as the ZX, but it's such a kick to drive." We concur and include this parting comment: At $9455 the Mazda is the sports/GT bargain of the era.

Porsche 924

HOW CAN a car that scored six category wins finish 3rd in cumulative standings? Simple. When it's good, it's very good, but when it's bad, watch out. Constant readers of R&T know that we've been complaining about the 924's problems (noise, lack of power) since the car's inception. The findings of this road ⟫→

PERFORMANCE / INTERIOR NOISE	Chevrolet Corvette	Datsun 280ZX	Mazda RX-7 GS	Porsche 924
Acceleration: time to speed, sec				
0–30 mph	2.5	3.8	3.8	4.3
0–60 mph	6.6	10.2	9.7	12.4
0–90 mph	15.0	24.9	22.3	29.7
Standing ¼ mile, sec	15.6	18.1	17.7	19.3
Speed @ ¼ mile, mph	91.0	80.0	82.5	75.5
Trip fuel economy, mpg	12.0	15.5	16.0	19.0
Braking: stopping distance, ft, from				
60 mph	142	160	151	150
80 mph	244	286	258	280
Control in panic stop	excellent	very good	very good	very good
Pedal effort for 0.5g stop, lb	25	18	22	25
Fade, % increase in effort in				
six 0.5g stops from 60 mph	nil	25	45	nil
Overall brake rating	excellent	very good	very good	very good
Cornering capability, g	0.798	0.760	0.779	0.766
Speed thru 700-ft slalom, mph	59.3	56.5	59.3	58.9
All noise readings in dBA:				
Idle in neutral	58	54	53	57
Maximum, 1st gear	84	81	80	86
Constant 30 mph	69	67	65	68
50 mph	73	69	71	73
70 mph	78	75	75	79
90 mph	83	78	80	83

test only reinforce our opinions. The Porsche was rated lowest in quietness and its engine finished last in its category by scoring only 29 points, the bottom of the numerical barrel. But let's look at the positive aspects of the Porsche, namely handling, braking, outward vision, exterior styling, exterior finish and interior styling. The 924 won all six categories, achieving its highest score in handling. Three drivers gave the car maximum points, two gave it 9, but tied it with the RX-7, and one gave it 7. "A road-hugger of the first order. Feels like it will stick and stick and stick," one said. "Handles and sticks like glue," said another. These statements reflect what we've said about the car in our road tests. It's a very neutral handling car with predictable attitudes. There's some trailing throttle oversteer, but an application of power tames it. With power on, the car has a tendency to understeer, but a bit more steering lock will make it turn. "Just keep applying more lock and more throttle and it keeps turning and tracking," one driver said.

Unfortunately, it's not always possible to apply more throttle because the 924's engine lacks adequate horsepower and is largely unresponsive except in a narrow band (4000 to 5000 rpm). There's no low-end power, a problem compounded by a rather sluggish automatic transmission. Downshifts are not instantaneous (the gearbox pauses before engaging), and this really becomes annoying when one attempts to shift the gearbox while cornering. And if sluggish performance is not enough, the 924's engine also offers noise—mechanical mutterings as well as a resonant buzz occurring at about 3200 rpm. As luck would have it, the engine reaches this speed and resonates at nor-mal U.S. cruising speeds of 55 to 60 mph.

In spite of our universal displeasure with the Porsche's engine and gearbox, we found the 924's brakes to be the best of the lot. The disc/drum combination showed no significant fade even though most of us used the brakes liberally while trying to make the 924 corner at its true (and quite respectable) potential. However, after one particularly twisty section the pedal developed an annoying pulsation and longer than normal travel. Our group liked the car's unassisted steering, though most drivers admitted it is heavy at low speeds, and there was that now-familiar problem with thigh/steering wheel interference when turning.

The 924's exterior styling, exterior finish and interior styling were rated best by our testers who also said that outward vision was another best-liked feature. There's not much more to say about the looks but we must point out that the 924's silver metallic paint is flawless. And the leather seats, an option on our test car, are luxurious to look at, feel and smell. Electrically operated side-view mirrors and window lifts are nice too. One last important plus—the 924 had the best fuel economy of all the cars at 19.0 mpg even though it had to be driven extremely hard to keep up with the three quicker cars. In fact, we found the latest average to be almost the same as the figure attained in our standard fuel economy test in which the cars are not driven nearly as aggressively. Our only conclusion is that regardless of the setting (twisty mountain road or around town), the 924 is a car that is nearly always driven practically flat out.

Good fuel economy, striking looks, a fine finish and the best handling in its field are pluses to be envied. But the question is, can they offset the 924's anemic performance, mind-numbing noisiness and high price?

Chevrolet Corvette

MANY OF us expected this car to fare much better than it did. A time proven engine with a sturdy transmission, reasonably good suspension enhanced by tires with a very large footprint—these are the things that made us optimistic about the Corvette. But after three days and more than 1000 miles, we could say only that the gearbox, heating and ventilation systems are first rate. The engine was rated 2nd behind the Mazda and it ranked 3rd in quietness, but it was last in the 16 other categories and accumulated only 850 points.

First, the good news: The gearbox, ventilation and heating categories are where the car excelled. There's not much to say about General Motors' Turbo Hydra-matic except to point out that it's strong, smooth as a baby's you-know-what and positive. We may not have liked the indecisiveness of the Porsche's gearbox, but we absolutely loved the assured-ness of the Vette's transmission. Moving the selector lever brings instantaneous response and no slippage. The irony of it all is that the Chevy V-8 has enough torque to absolutely minimize down-shifting in corners, except under the most extreme conditions.

The Corvette's heating and ventilation systems are among the best in the industry. With the flick of a lever one can have as much heat or chilled air as desired. Unfortunately, there's more to a sports/GT car than climatic bliss. There's handling, ride, driving position, seat comfort and body structure, just a few of the areas where the Corvette did poorly. Take handling. Our skidpad figures and slalom speeds will tell you that the Vette performs admirably. It corners flat, especially when equipped with the gymkhana suspension option which our car didn't have. There's a general tendency toward understeer, the safety device of the inexperienced driver. And just for luck, plenty of rubber makes the car stick in almost any attitude. But on our trip, most drivers never had the opportunity to take advantage of the car's inherently good handling. They were thrown off balance by the steering which has initial quickness but then slows down to keep the average driver from turning too sharply. One driver said "You need a delicate touch and discipline to keep from cranking in too much lock." But another, who said he was "always fighting with the steering," admitted he didn't know just how much lock to crank in. In this case, unfamiliarity bred contempt.

The Corvette's harsh ride and loose body structure gave the car low marks in two more categories and contributed to what by now was considered poor handling. The tires thumped, the body rat-

Datsun's dual-range fuel gauge is an example of the car's top-rated instrumentation.

The middle switch of the three window-lift controls offers one-touch opening and closing of the Datsun's left side window.

CUMULATIVE RATING SHEET

	Chevrolet Corvette	Datsun 280ZX	Mazda RX-7 GS	Porsche 924
Performance:				
Engine	50	47	**55**	29
Gearbox	**53**	49	50	35
Braking	48	49	50	**53**
Steering	33	43	**54**	46
Ride	31	**54**	48	45
Handling	39	45	54	**55**
Body structure	31	**53**	45	45
Comfort/Controls/Convenience:				
Driving position	38	52	**54**	39
Seat comfort	37	**49**	47	45
Controls	40	**55**	53	46
Instrumentation	45	**54**	49	50
Heating	**54**	52	51	44
Ventilation	**52**	48	42	38
Luggage space & loading	29	**54**	45	50
Outward vision	44	46	49	**53**
Quietness	36	**53**	48	35
Design/Styling:				
Exterior styling	37	42	45	**53**
Exterior finish	38	52	53	**54**
Interior styling	39	**49**	45	**49**
Interior finish	43	**56**	52	53
Ingress & egress	38	**52**	44	40
TOTALS	855	**1054**	1033	957

tled and on rough roads the car was skittish. "When the going gets rough, this car feels every ounce of its 3655 lb," quipped one driver. These aforementioned ills caused two drivers to knock points off the brake scores. However, in our objective brake tests the Vette was a clear winner.

Relative to the other three cars, the Corvette's 11-year-old design contributed to its low scores in exterior and interior styling and instrumentation. Outward vision, in part a result of the Vette's styling, was another least-liked area. But few things were as vehemently criticized as the Corvette's driving position, seating comfort and luggage space. The driving position is simply uncomfortable. The tilting/telescoping steering wheel is a plus, but the seats are poorly shaped, offer few essential adjustments (rake, for instance) and are too narrow. Two drivers complained of neck cramps after only a half-hour of driving. But the upholstery does a decent job of keeping occupants in their seats.

So far, so bad. The Corvette's cargo area came in for some well directed criticism. Although it's reasonably large (9.5 cu ft), the rear deck is accessible only through the passenger compartment after one folds down the seat. True, the new clamshell seat design improves access, but loading luggage through the passenger compartment is still not a laudable concept.

Up to this point we have said little about the Corvette's engine except that it has lots of torque. Here are the details. In California trim, the powerplant is a bit disappointing, because it runs out of breath at about 4500 rpm and shows a reluctance to rev past 5000. It also surges a bit and detonates if fed a nutritionally deficient diet of unleaded gasoline (bad fuel). And it is thirsty. In just over 1000 miles of driving, fuel consumption averaged 12.0 mpg.

Now here's the bright side. The Chevrolet V-8 is quiet, even at speeds up to 90 mph. It's still the fastest in a straight line and has gobs of torque, a very important attribute for most American drivers. Mechanically simple, it's the sort of engine you can forget about and yet still rely on.

Conclusion

IT WOULD be difficult not to choose the RX-7 or the 280ZX when buying one of these sports/GT cars. Both made a fine showing and serve to remind us that the Japanese are serious about their products. The ZX, which lacks the crisp handling the 280Z had, now becomes a luxury GT, just the way Nissan intended. The RX-7 steps in to take the old Z's place and becomes a sports car bargain even at a price of nearly $10,000 including options. The Porsche 924 offers the best handling of the lot, a crisp Teutonic line and a sort-of understated Old World quality. But its lackluster performance adds insult to injury by suggesting that someone pay $18,000 just to be humiliated on the highway. Finally, there's the Corvette, the only U.S.-built sports car. Much loved and still very desirable, it's quick, has excellent brakes, a superb automatic transmission and is filled with many appreciated amenities such as cruise control, electric window lifts and a tilt/telescope steering wheel, all of which make it an excellent value.

That's how we rate these four automatic-equipped sports/GT cars. But before you make a choice, read Jim Crow's story on the rating system. You may find that it takes more than numbers to find that indefinable quality that makes one car just right for you.

REMINISCENCES & REFLECTIONS

I USED TO work full-time at *Road & Track*. I was there, for instance, when the first multi-car comparison test was done. It compared 10 imported station wagons and appeared in the June 1968 issue. What I remember most vividly about it was the aggravation involved in seeing how many of what size suitcases could be jigsaw-fit into each luggage area.

Having been involved in that historic event, I was pleased to be invited to participate in this most recent comparison. I found it interesting to learn how much—and how little—the procedures have changed in the intervening years.

Most of all, I think I was impressed by how much faster/harder the current staff drives the cars. It seems to me we used to have only one driver who was content only when driving at the absolute maximum. Now they all drive that way. All the time. Or at least all the time when it is prudent to do so. Given an open, empty road, be it straight or corkscrewed, there was an abundance of 10/10ths motoring.

For the purposes of this test, I think this is good. The route covered by the four cars totaled 1020 miles, less than a hundred of which was over the ordained dullness of interstate highways. The intent was to let the cars show what they were capable of doing. There were long, straight stretches where it was possible to place the pedal against the firewall and leave it there mile after mile. There were other sections that thoroughly tested not only the car's ability to stay on the road but also challenged the driver's skill in keeping it there. And the distant glimpse of a suspicious silhouette revealed volumes about the vehicles' ability to decelerate without melodrama.

How much do you learn from this kind of motoring? Although you could make a case for the results being misleading since almost no one is able to drive like this in the real world, these are nevertheless the criteria by which such cars should be measured; indeed must be measured so long as the evaluation of such cars has any validity at all. If you insist on applying only those standards pertaining to the world of the legislated 55 max, there's no reason for any car to offer even as much performance as your basic 4-cylinder Rabbit. And no enthusiast is ready to settle for that.

Also, in my return to the world of comparison tests, I found that the scoring system has changed, although this is one of degree rather than basic procedure. When we first used score sheets in our evaluation (June 1970, "Four Sports Cars") each car was rated 1-to-4 for "Best" to "Worst" in 14 different categories, then they were rated in order of overall preference, irrespective of individual scores. Now there are 21 categories (inflation?) and scoring is on a 10-possible basis where each characteristic is assigned a value between 1 and 10. And there is also a rating for overall preference, as before.

How much does the scoring of categories tell you? Personally, I doubt the necessity for such fine-line nitpicking and that the system is imperfect is shown by the fact that the totals for various vehicles may not reflect the drivers' overall preferences.

But this isn't really serious since such anomalies cannot be avoided in any system simple enough to be practical. To achieve scoring perfection it would be necessary to weigh each category, deciding the relative value of each characteristic in relation to every other. Such things as engine, gearbox, brakes and handling would heavily outweigh such non-functional characteristics as ingress/egress, styling, finish and so on. Establishing the values for such a scoring system would probably prove impossible—especially if all six drivers were asked to agree on the weighted value given to each.

And all this probably wouldn't be worth the trouble anyway since the driver's overall preference is the single most important result of all the effort that goes into such a test.

Overall, I have no doubt about the validity of the results. What a comparison test such as this provides the reader is the opinion of six drivers not unlike himself who were able to do what any driver would like to do before making up his mind about what car to buy.

And you can't ask much more than that.—*James T. Crow*

ANOTHER PUFFIN' PORSCHE

*This one undercuts
the factory 924 Turbo in performance and price*

BY JOHN DINKEL

YOU MIGHT THINK R&T is starting to sound like a broken record, but damn it, we're frustrated. The cause of that frustration? Porsche's 924. On paper it has all the makings of a great GT. In truth, the steering, handling, interior layout, seating, instrumentation and styling are all in the Porsche tradition. And we must admit that evolutionary improvements have alleviated a few of the niggling problems that have plagued the 924 since it was introduced in 1976 (especially a classic case of *freeway hoppus Californium*). This brings us to the reason of our discontent: the engine. The Audi-designed 2-liter is noisy, rough and buzzy and, adding injury to insult, gives performance hardly commensurate with a GT of the 924's supposed stature and price—now up to $14,600.

Boast about an 0–60 mph time of 11.0 seconds or running the quarter mile in 18 sec flat to your less well-to-do friends in their Toyota Celicas, VW Sciroccos and even Dodge Colt Hatchbacks and VW Rabbits and the laughter will be heard clear to Stuttgart. And short of committing hara-kiri (that only works with underpowered Japanese cars), your next move might be to slow-tail it in the direction of your local Porsche purveyor and hand the

salesman a deposit on a 924 Turbo. If you're well heeled, the Turbo's expected $20,000 price tag won't dissuade you from using such an extreme method as a means of gaining back a measure of your wounded pride. But if you're like most 924 owners we know, you purchased a 924 because, one, it's a Porsche and, two, it's relatively affordable. And a twenty-grand price tag is enough to make you contemplate trading in your string-back driving gloves for some insulated mittens and grabbing the next slow boat to Alaska.

Keep the faith, 924 owners, there's another solution. Have you thought about aftermarket turbocharging? Yes, I know you've heard all the turbocharger horror stories: blown head gaskets, pieces that don't fit, instructions so complex and obscure you need PhDs in engineering and cryptography to decipher them, blown head gaskets, overheating, detonation, blown head gaskets, exhaust leaks, blown head gaskets.

Obviously there's some truth to the stories; sometimes the kits are improperly designed, but more often than not, it's careless installation or driver abuse that ultimately results in problems. There's not much I can do about the latter, but I can tell you about an aftermarket turbo for the 924 that transforms the Porsche into what I believe is the most flexible, most driveable and nicest running turbocharged car—production or after-market—I have driven. Admittedly, I haven't tried the factory's 924 Turbo, which Joe Rusz found is a sensational performer in European trim (R&T February 1979), but I've driven and tested the production turbocharged Ford Mustang, Buick LeSabre and Riviera, Saab Turbo, Porsche 930 Turbo and several aftermarket conversions. As far as I'm concerned, none can match the 924 turbo system developed by Windblown Systems located at 158 Merrick Rd, Amityville, N.Y. 11701; 516 691-1733.

Fred Dellis, President of Legend Motors, a Porsche-Audi and Fiat dealer in Amityville and President of Windblown Systems, emphasizes that "system" not "kit" is the proper way to describe the turbo Windblown has engineered for the 924. "It's a system," Dellis says, "because it's a totally professional execution of engineering, manufacturing and testing down to the smallest detail. It's an excellent complement to the fine product it is to be used with, using the same standards, or higher, of workmanship and quality throughout."

boost gauge. Consider attention to detail. Check the photo showing the system in exploded form. Everything you need to install the turbo properly—down to the smallest washer, cap screw or hose clamp—is provided. You'll find heat shields for the sparkplugs and the alternator drive belt and a compatibly engineered exhaust system of 2.5-in. diameter mild steel tubing, including a muffler. You also get a clearly written 10-page instruction manual complete with diagrams and a template for correctly drilling the oil pan.

Of course, it's results that count. The system must perform and also—and this is especially critical for a turbocharger installa-tion—it must be durable and reliable. Durability and reliability of the turbocharger and the engine it's attached to are functions of several factors: how well the kit or system is engineered, proper installation, regular maintenance and driver sensitivity. Ob-viously, starting a cold engine and immediately giving it a dose of full boost is not conducive to longevity. To date the total mileage on the four turbo 924s Windblown Systems owns exceeds 40,000 miles, with one car accounting for 25,000 of those miles including 54 hard laps around Bridgehampton Raceway. Dellis reports no

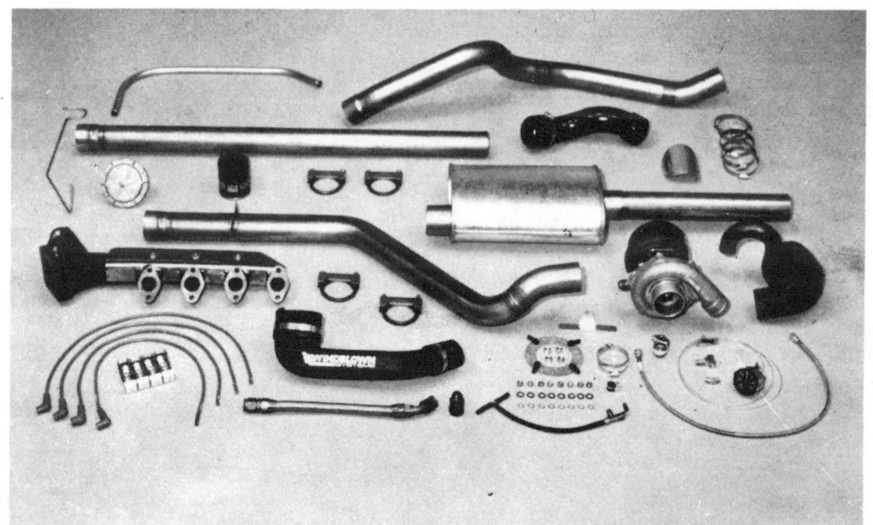

Not just a kit, but a system. Windblown's package includes all necessary bits (left), plus VDO boost pressure gauge (top).

PHOTOS BY THE AUTHOR & LESLIE DINKEL

Words like those are easily spoken, but Dellis can back them up with facts. Consider quality. Windblown provides an AiResearch turbo, a special cast iron alloy exhaust manifold, a cast iron turbine discharge housing, an aluminum compressor discharge casting, Aeroquip braided stainless steel oil feed and drain lines, silicone sparkplug wires, a Rotomaster wastegate and a VDO

problems or failure with any of the cars.

Results are easier to quantify. My testing of a 1978 4-speed equipped turbo 924 resulted in 0–60 mph times averaging 6.7 sec and the quarter mile flashing by in 15.3 sec at 90.5 mph, reductions of 4.3 and 2.6 sec, respectively, compared to a normally aspirated 924. Want more proof? The time to reach 90 mph is nearly halved: 15.2 sec vs 27.2. That's impressive. But if you're familiar with turbo installations you're aware that the gain in top-end performance often comes at the expense of bottom-end acceleration. Lay your fears to rest. The Windblown turbo scoots from 0–30 mph in 2.1 sec compared to 3.3 for the stock 924. And that time is only 0.2 sec slower than a 930 Turbo.

This ability to have your cake and eat it too is the result of a major breakthrough in turbocharger design—bias boost control. It works this way: A pressure signal from the exhaust upstream of the wastegate and a vacuum signal from the intake manifold are fed together in a predetermined ratio directly into the wastegate. The pressure/vacuum ratio the wastegate actually "sees" is a function of the velocity of the exhaust gases and the depression in the intake manifold. Both of these, in turn, are a function of throttle opening and engine rpm. Tricking the wastegate in this fashion means that with an 11-psi relief spring in the wastegate, the turbocharger produces 9 psi of boost from 2800–4800 rpm; from 4800 to redline the boost progressively drops to 5 psi.

This does several good things. Restricting the boost as the load, the rpm and thus the peak cylinder pressures and temperatures build up, matches the boost curve of the turbo to the engine's octane requirements. With bias boost a Windblown 924 can ⟫⟫→

operate without detonation on generally available 93-octane premium unleaded fuels. A conventional wastegate set to blow off at 7 psi wouldn't give the optimum safe 5-psi boost Dellis feels is necessary at high rpm. Neither would it allow the extra low-end punch achieved by going to 9 psi boost at low rpm when the engine is less susceptible to detonation. It also means the stock 8.5:1 compression ratio can be retained. This is one of the reasons why the European factory 924 Turbo engine has a 7.5:1 compression ratio with its 10 psi of boost; under full boost the effective compression ratio jumps to 10.8:1.

Only one word aptly describes the Windblown turbo 924 installation: sanitary. The turbo nestles in the right front of the engine compartment with the wastegate directly above it. The precision look of the exhaust manifold casting is particularly impressive. In fact, the whole installation is so professional looking you could be fooled into thinking it was engineered in Stuttgart, except for the Windblown Systems, Inc lettering on the handsome aluminum compressor outlet casting that mates with the stock throttle housing.

A few words about the wastegate. The Rotomaster unit regulates the turbocharger speed by controlling the volume of the turbine inlet gases, a design that allows the turbocharger to produce only the required boost level without limiting the fuel/air flow through the compressor stage of the turbo. Such a design would cause the turbo to operate less efficiently, resulting in higher intake manifold temperatures and increasing the possibility of detonation.

While the wide-open-throttle performance of the Windblown 924 is certainly sensational, what impressed me even more was the driveability and flexibility of the engine under all driving conditions. Turbo lag was almost unnoticeable; I could have sworn the engine was normally aspirated if I hadn't had a boost gauge to watch. That quick 2.1-sec 0–30-mph time is proof that this turbo is highly motivated. But I could stomp on the throttle with the transmission in 4th and the engine trundling along at 1500 rpm and immediately feel the engine smoothly build power. Try that with a stock 924 without shifting down at least two gears and you'd find yourself going nowhere slow.

Deceptive is how I describe the turbo 924's performance. There's no neck-snapping acceleration such as you experience with a 930 Turbo. Instead the smooth, quiet power buildup lulls you into disbelieving the speedometer, which in these days of 55-mph speed limits can be a costly mistake. There's nothing deceptive about what the turbo does to the 924 engine's usual assortment of rattles, wheezes and booming resonances: It muffles all of them. In addition, the engine runs through its rev range so quickly that vibrations, though still present, don't have time to jangle your nerves.

Dellis believes the only way to support a high quality product like the Windblown turbo 924 system is with a well structured distributor network supported by high-caliber factory sales and service personnel. Such a network is currently being developed. Also underway is the testing necessary to obtain a clean bill of health from the Environmental Protection Agency. Windblown Systems has submitted a 924 for testing under federal emission test procedures and based on the results of the test, indications are that conformity has been met. By the time this issue appears, written conformation should have been obtained. And work is proceeding on similar turbo systems for BMW 320i, 528i, 633CSi models, the VW Scirocco and the Fiat X1/9. Compared to most turbo kits, the Windblown system is expensive: $2085. Installation—averaging 18–20 hours for the first-time installer—would cost an extra $350–$400. On the other side of the ledger, however, are the quality and completeness of the system, the impressive performance gain per dollar spent, the enormous improvement in flexibility and the welcome reduction in engine noise and roughness. Any current or prospective 924 owner who can't afford a factory Turbo owes it to himself to find out more about the Windblown turbo. The transformation truly has to be driven to be believed.

Cargo area, which can be augmented by folding the rear seatback, is hidden from view through huge window by a plastic cover that rolls up like a shade when not being used.

Audi-built, fuel-injected engine develops 110 bhp.

CONTINUED FROM PAGE 6

the touches one expects from a driver's car—nay, a Porsche. Gripes? Only that the asymmetrically centered steering wheel hits the thighs when turned more than 90 degrees. Porsche could remedy this by raising the steering column or lowering the seat, but then the wheel would be too high. It is a fundamental problem of the 924's interior package.

A Porsche's strong point has always been handling and the 924 lives up to that tradition, at least with the optional anti-roll bars. It is a car that the novice can drive quickly and with confidence. Weight is distributed nearly equally front to rear (49/51 percent), so the 924 is quite neutral. Its normal characteristic is mild understeer which can be controlled with judicious application of the throttle. The car can be pitched about, yet is forgiving enough to keep most drivers out of trouble. The tires howl and the body leans, but on the skidpad the 924 pulls 0.766g lateral acceleration while managing almost 59 mph in the slalom. That's better than the 280Z and the Alfetta GT, the 924's major competitors.

At almost $12,000 the 924 is far from a bargain. Blame it on the devalued dollar, but be prepared to bite the financial bullet because unless you're willing to settle for an absolutely plain-Jane Porsche you'll have to pay extra. Although one could easily tack another $2000 to $3000 (or more!) onto the base price of a 924 by ordering capriciously, a judicious shopper can outfit his car properly by ordering one of the many touring, convenience or sport packages. Each package offers its own options which include such items as radio, radio speakers, antenna, alloy wheels, leather-covered steering wheel and anti-roll bars. Prices range from $520 for Touring Package I to $1270 for Convenience Package II, whose hefty tab includes Porsche's best AM/FM/shortwave signal-seeking radio with cassette tape player as well as power windows.

It's also possible to buy many of these add-ons individually, and because the list is extensive we'll tick off just the most viable ones: front and rear anti-roll bars, $150; rear-window wiper, $185; Bilstein shock absorbers, $135; alloy wheels, $390 ($470 in black) and leather-covered steering wheel, $120. From there on it's onward and upward with such things as air conditioning ($595), removable sunroof ($380), power windows ($295), leather seats ($635) and metallic paint ($395) to make your 924 a reflection of your personality—and your net worth.

The long-awaited 5-speed standard shift gearbox is not presently available but a 3-speed automatic, costing $399, is. That's not very encouraging news for the enthusiast, but it is an appealing idea for the in-town driver. Don't expect great things in terms of performance, though.

What's to be expected from the 924 in the future? Performance commensurate with its looks and handling, quite probably. As already mentioned, turbocharging is rumored. So is a new powerplant: The alternatives mentioned include the Audi 5-cylinder (which Porsche insists is too long), a yet-to-be-seen V-6 and a four that's essentially half the 928 V-8. Not that it allays the 924's current faults, but the 924 probably does have a fascinating evolution ahead of it.

PORSCHES ON ROAD & TRACK

Porsche performance and R&T testing have grown together

BY JONATHAN THOMPSON　　　　　　　　ILLUSTRATION BY WM. A. MOTTA

IN THIRTY-TWO YEARS, and not quite 30 volumes, *Road & Track* has published 1000 road tests, of which 46 have been on Porsche automobiles from the 356-4 in November 1952 to the 924 Turbo in this issue. Actually, 53 individual Porsches have been involved in these tests, which have included group, comparison, update and European evaluations in addition to those conducted according to the standard format. But even the term "standard format" refers to a steadily evolving procedure in which the magazine presentation as well as the test methods have changed. For this reason, it is not possible to make absolutely accurate comparisons of earlier Porsche test results with those obtained today, but a survey of these tests makes an interesting pattern, in which Porsche and *Road & Track* have both grown in complexity and sophistication.

While the 46 tests comprise every type of Porsche from stock to racing, and engines of four, six and eight cylinders, we will concentrate in this survey on the "basic" 4-cylinder lineage. Even here we are really talking about a number of different engines, as well as three different chassis configurations: the original, now classic (Volkswagen and) Porsche rear layout, the mid-position unit of the 914 and, of course, the current front-engine, rear-drive of the 924. Ironically, it is this present configuration which used to be called "classic," now given new life by many manufacturers who have found it still the most efficient for overall performance, especially when habitability and cost are priorities.

But the original Porsche was a rear-engine, air-cooled, VW-based car, and in the early years the Wolfsburg connection was no cause for scorn. Humble beginnings are expected of all great legends (even the first Ferrari, the Vettura 815, employed Fiat engine components) and it was only later that "purists" questioned the ingredients necessary for a true Porsche, making the 912, 914 and even the 924 something less than thoroughbreds in certain circles. But the fact is that the name Porsche almost insures that the car in question will find an eager clientele, at a price usually exceeding those of otherwise direct competitors.

The 356-4 was the original, inverted-bathtub type Porsche four, with 1488 cc and 65 bhp. It was really a GT car and its streamlining, unusual at the time and not considered attractive by everyone, contributed to the car's great efficiency: 103 mph and 35 mpg. In those days Porsches had the kind of handling to be expected from a VW-based, tail-heavy car (it certainly wouldn't feel comfortable to a driver used to sophisticated Seventies handling), but it was more than good enough for enthusiastic sporting use at the time. The $4208 price seems tiny today, but it was a lot of money in 1952.

A number of similar 356 tests followed, with engine and body changes making the cars significantly different. The Super coupe tested in 1954 had 70 bhp and went 108 mph, while the first open Porsche tested was the Speedster, in 1955. This was a 1500S with 84 bhp; with its slightly lighter weight, really good acceleration was achieved (0–60 in 10.3 seconds would satisfy a lot of Porsche owners today) although lesser aerodynamic efficiency lowered the maximum speed to just over 100 mph with the top down. The Continentals tested together in January 1956 were not remarkable for their performances, but their prices were enticing to say the least: $3590 for the Coupe and only $2995 for the Speedster! Performance *was* the forte of the Carrera, with a 1498-cc, 4-cam, 550-type engine producing 110 bhp and giving the Coupe a top speed of 120 mph. Of course it wasn't cheap, at $5995 in 1956. Suspension was modified to keep pace with the available performance, on the other models as well as the Carrera, and the trend of constant Porsche refinement of a basic product was becoming evident. Although not part of the regular production program, the 550 Spyder was tested in February 1957. This mid-engine racing car, made famous by Ken Miles among others, had 137 bhp and, weighing only 1510 lb, accelerated impressively: 0–60 in 8.2 sec and the quarter mile in 16.1.

Still carrying the original 356 bodywork, the 1600 Coupe, 1600

Super Speedster and 1600 Convertible D brought increased options and sophistication to the Porsche buyer from 1957-1959. The 70-bhp Coupe was sedate in performance but a nice GT car at $3790, the Super Speedster used its 88 bhp to do 0-60 in 10.5 sec for $3928, and the Convertible D fell between the two in purpose and appeal: an open car with less performance but more comfort than the Speedster for $3695.

Tested in March 1960, the Super 90 was a Speedster to the new 356B configuration, with revised bodywork and a 1582-cc, 102-bhp version of the pushrod four. Performance was a nice balance of acceleration (0-60 in 12.5) and top speed (117 mph), while the improved interior and other details were a bargain for $4195. The 1600 Normal (October 1961, the first Porsche test with a full-page data panel) was a modest performer while the Super 90 tested in the same issue was a 102-bhp short-course competition car with good acceleration but a maximum speed of only 89 mph (115 was the maximum with 3.78:1 gearing).

Tested in Europe, the Carrera 2-liter of July 1962 had no less than 152 bhp, got to 60 mph in 9.2 sec and topped out at 123 mph. By this time Porsches had become recognized as a special breed of automobile and the price of $7595 was just part of the Zuffenhausen experience. Nearly two years were to go by before the next Porsche test, of a 1582-cc, 88-bhp 356C; this was a 100-mph automobile for $4295. My first direct contribution to R&T came in July 1964 in conjunction with the next Porsche test, a set of scale drawings of the 904 GTS published with the regular report. The 904, a mid-engine, fiberglass coupe weighing only 1350 lb dry, was a famous GT competitor raced all over the world by the factory and many private entrants; if you were in the latter category, the base price was $7425, but it goes without saying that a fully race-worthy 904 cost its owner a bit more than that. As tested, the GTS got to 60 in 6.4 sec, the quarter mile in 14.5, and peaked out at an estimated 150 mph. Skidpad and slalom tests were not done by R&T in those days but the figures would have been impressive, taking into account the 15-in. Dunlop rubber employed.

The era of the 6-cylinder Porsche road car began with the 2-liter 901, later 911, and our first test was published in March 1965. This $6500 coupe, with a body design still used today, started Porsche's move from the medium into the upper price bracket. Incredibly, the 911 is the basis for the 935 Turbos which are, according to how you look at them, either 911s with 917 expertise incorporated or 917s in 911 clothing, getting as much performance from six cylinders as the non-turbocharged twelve used to put out. With 145 bhp and weighing 2360 lb, the first production 911 was a good if not startling performer (0–60 in 9.0 sec, 132 mph in 5th gear) but the emphasis was on an overall blend of performance, handling, comfort and detail quality that has been the mark of the top-line Porsches ever since. R&T went on testing many more 911 and 911-derived sixes (see accompanying chart), right up to the ultimate, the Martini factory-run 935 Group 5 Turbo. This car did not have an acceleration *curve;* it was a straight line, hitting 60 mph at 3.3 sec and going off the graph at 150 mph and 11.0 flat. The maximum speed, dependent of course on gearing for a particular circuit, was 183 mph on the example tested, while the lateral acceleration was estimated at a boggling 1.400g. Price? Well, it just *started* at $75,000.

But back to the fours. Beginning in 1965 (our test was in February 1966) the 912, essentially a 911 with the older 1582-cc unit, brought the newly achieved level of comfort and handling, but not the performance, to buyers with $4700 to spend. In effect, it was the first "junior" Porsche, a distinction that has been made among owners, if not by the factory, in the model range ever since. The 912 was no slouch, though: 0–60 in 11.6 sec and a top of 119 mph were just fine for a lot of buyers.

The next part of the Porsche 4-cylinder story is a curious one, in that the factory was not ultimately successful and the whole project seemed compromised from the start by conflicting marketing intentions. Called the VW-Porsche (correctly) in Europe, the mid-engine 914 was the next-generation junior Porsche over here; loved by some, scorned by others, it never managed to make its supposedly advanced chassis configuration as satisfying as it was promising. The main trouble was the homely styling (when I visited the factory for its 1969 unveiling, my first reaction was, "This is a Porsche?") while the noisy VW powerplant was not what a lot of Porsche enthusiasts were seeking. But the 914, first tested by R&T in April 1970, came in at $3595 initially and certainly brought in a new group of buyers, many of them undoubtedly VW owners who had always wanted to move up and who didn't object to their familiar VW four being put to sporting

A sampling of 4-cylinder Porsches. The 356 Super (above) tested in September 1954, two 1500 Continentals, Speedster and Coupe (above right) road tested in January 1956; Super 90 (below right) evaluated in March 1960; Type 356B (bottom right) introduced in 1959 with enlarged rear window compared to previous model; the 1600 Super Speedster tested in April 1958 was the first of the 356s without roller-type engine bearings.

use. With 85 bhp from the 1679-cc unit, it did well to reach 60 in 13.9 sec and 109 in 5th.

The 914 was tested again in June of the same year, in comparison with the Fiat 124 Spider, Triumph TR6 and MGB Mk II. The Porsche was rated second to the Fiat in overall appeal. For those who insisted on a Porsche engine and Porsche performance from a Porsche, the 914/6, tested in the next issue, made the 911 six available for a brief production run. This car cost $6099 and is a collector's item today; performance was very good (0-60 in 8.7 sec, maximum speed 123 mph) but our consensus was that the contemporary 911T was a far better machine for only $431 more. To get more out of the standard four, Porsche produced the 914 2-liter (tested in February 1973), which made a significant improvement in performance over the original (0-60 in 10.3, 119-mph top) but by then the base price was $5299. Skidpad performance was 0.742g, very good but not noticeably superior to the 911E, which benefited from years of chassis refinement. The 1.7 version of the 914 was track-tested in April 1973 along with eight other showroom-stock candidates; best in braking, the 914 came in third in most categories, including lap times, in which it tied the MGB, behind the Triumph GT6 and Opel GT. The advantages of the mid-engine layout (or lack of them) were explored in a July 1973 test which confronted the 914 with an Opel GT (traditional front-engine, rear-drive), a Saab Sonett (front-drive) and a VW Karmann Ghia (rear-engine). The 914 was fastest in slalom and lane-change tests, as expected, while serviceability and space utilization were not ideal. In fairness, the last factor is not critical in a performance-oriented 2-seater, but by this time even Porsche was thinking of abandoning the mid-engine layout for GT production cars.

In July 1976 the first of the Audi-related Porsches, the 924, was tested in comparison with the Alfetta GT and Datsun 280Z. R&T rated the 924 a winner over its Italian and Japanese competition, but some old-hand Porschists could not regard a front-engine, water-cooled car as a member of the family (the V-8 Porsche 928 has helped dispel that attitude since then). But the 924 avoided the stigma of the previous junior Porsche, the 914, by being a far better-looking automobile if nothing else, and the car was more refined in other ways as well. All this for a price which emphasized the escalated German Mark; even at the "low" $9400 price ⟩⟩⟩→

The last 356 model (top). Starting in 1949 Porsche produced 76,302 cars with the 356 body. The 904 (above) from July 1964 did 0-60 mph in 6.4 sec. The 914 (left) was the first affordable mid-engine car. The 912 (below) had the 911 body wrapped around the 1582-cc 4-cylinder.

estimated at the time, the 924 cost nearly $3000 more than the Datsun and $1000 more than the Alfa. Although its 95 bhp did not produce startling performance (0–60 in 11.9, maximum 111 mph), its competitors couldn't match the handling.

An update of the 924 was published in August 1977. At a still-low $9995, it had 15 additional bhp that were put to good use, giving 0–60 in 11.0 sec and a top of 117 mph. The ride had improved over that criticized in 1976, but the noisy engine was not found worthy of a Porsche. In April 1979 an automatic 924 was compared with versions of the Corvette, Datsun 280ZX and Mazda RX-7 using similar transmissions. Unfortunately, the Porsche was the slowest and the most expensive of the group and we said, "Its lackluster performance adds insult to injury by suggesting that someone pay $18,000 just to be humiliated on the highway." What the 924 lacks is power.

Now we come to the 924 Turbo tested in this issue. It is evident that Porsche has endeavored to restore its performance image in the 4-cylinder, junior category. The 924 Turbo is certainly not junior to many cars in price (an incredible $21,000 estimated) or performance. With 143 bhp from the turbocharged 1984-cc four, this 2835-lb 924 can reach 60 mph in 7.7 sec, the quarter mile in 16.3 and a maximum of 132 mph in 5th.

It is clear that the development of the 911 series (and its 935 racing derivatives) must come to an end some day, and that concentration on the 924 body/chassis package for future high-performance Porsches will take place. We are likely to see the same dedicated, year-by-year refinement of the basic design that marked the history of the 911 line, and the development of specialized competition-oriented 924s for SCCA club racing as well as international events. It goes without saying that we will be testing them.

In 30 years Porsches have increased from fourfold to tenfold in price, depending on the models considered. Half of the increase has been simple inflation, mostly in the last six years and compounded by the car's German origin, while the rest can be ascribed to constantly improving technology and driving sophistication. *Road & Track* has gone up sixfold in cover price during the same time span, while increasing nearly sixfold in average number of pages, and we like to think we're packing in a lot more performance and sophistication, too.

PORSCHE ROAD TESTS IN ROAD & TRACK

356-4 Coupe, November 1952
Super Coupe, September 1954
1500S Speedster, May 1955
Continental Coupe &
   Speedster, January 1956
Carrera Coupe, September 1956
550 Spyder, February 1957
1600 Coupe, August 1957
1600 Super Speedster, April 1958
1600 Convertible D, April 1959
Super 90 Speedster, March 1960
1600 Normal Coupe, October 1961
Super 90 Competition, October 1961
Carrera 2-Liter Coupe, July 1962
356-C Coupe, February 1964
904 GTS, July 1964
911 Coupe, March 1965

912 Coupe, February 1966
911S Coupe, April 1967
911 Sportomatic Coupe, February 1968
911E Coupe, January 1969
911S 2.2 Coupe, March 1970
914, April 1970
914 (comparison), June 1970
914/6, July 1970
914/8 (Chevy V-8), December 1970
911T Sportomatic Coupe, April 1971
914 (update), February 1972
911E 2.4 Coupe, February 1972
914 2.0, February 1973
914 1.7 (comparison), April 1973
914 2.0 (comparison), July 1973
911S Coupe & Carrera RSR, August
   1973

911, 911S & Carrera Coupes, January
   1974
911 Targa (comparison), February 1974
911 Carrera RS, October 1974
911 Carrera 2.7, March 1975
912E, 911S & 930
   Turbo Carrera Coupe, January 1976
924 (comparison), July 1976
Turbo Carreras: production,
   Groups 4 & 5, January 1977
924, August 1977
928, April 1978
911SC Coupe, April 1978
930 Turbo Coupe, June 1978
928 Automatic, December 1978
924 Automatic (comparison), April 1979
924 Turbo, June 1979

SIGNIFICANT PORSCHES VS CONTEMPORARIES

	Year	List price	Curb weight, lb	Wheelbase, in.	Engine, cc	Bhp @ rpm	0–60 mph, sec	0–1320 ft (¼ mi), sec	Maximum speed, mph
Porsche 356-4	1952	$4208	1850	82.7	1488	65 @ 4800	13.8	18.4	103
Jaguar XK-120	1951	$3495	2750	102.0	3442	160 @ 5400	10.1	18.3	122
MG TD	1953	$2157	2005	94.0	1250	54 @ 5200	19.4	21.3	79
Ferrari 212	1952	$9500	1975	88.5	2562	170 @ 7200	7.0	15.3	123
Porsche Super 90	1960	$4195	1935	82.7	1582	102 @ 5500	12.5	17.7	117
Lotus Elite	1960	$5244	1420	88.0	1220	76 @ 6100	12.2	18.0	115
Fiat 1500 Sport	1960	$3812	2200	92.1	1491	90 @ 6000	10.6	18.5	105
Alfa Giulietta Super Spider	1959	$3887	2040	86.6	1290	103 @ 6000	11.0	17.1	113
Porsche 912	1966	$4700	2100	87.0	1582	102 @ 5800	11.6	18.1	119
BMW 2000CS	1966	$5185	2630	100.4	1990	135 @ 5800	11.3	18.2	115
Volvo 1800S	1966	$4200	2410	96.5	1780	115 @ 6000	13.9	19.0	109
Alfa Giulia Sprint Veloce	1965	$3595	2150	88.6	1570	129 @ 6500	10.5	17.4	109
Porsche 911S	1967	$7074	2365	87.0	1991	180 @ 6600	8.1	15.7	141
Shelby GT 500	1967	$5114	3520	108.0	7014	355 @ 5400	7.2	15.5	132
Sunbeam Tiger II	1967	$3842	2560	86.0	4737	200 @ 4400	7.5	16.0	122
Ferrari 275 GTS	1966	$14,500	2960	94.5	3286	260 @ 7000	7.2	15.7	145
Porsche 914	1970	$3595	2085	96.4	1679	85 @ 4900	13.9	19.2	109
Datsun 240Z	1970	$3525	2355	90.7	2393	150 @ 6000	8.7	17.1	122
Fiat 124 Spider	1970	$3450	2090	89.8	1438	96 @ 6500	11.9	18.3	104
Triumph TR6	1970	$3425	2360	88.0	2498	104 @ 4500	10.7	17.9	109
Porsche 924	1978	$11,995	2575	94.5	1984	110 @ 5750	11.0	18.0	117
Alfa Romeo Sprint Veloce	1978	$9395	2660	94.5	1962	111 @ 5000	10.1	17.8	116
Datsun 280ZX	1978	$9500	2825	91.3	2753	135 @ 5200	9.2	17.2	121
Mazda RX-7	1978	$7275	2420	95.3	1146 (rotary)	100 @ 6000	9.2	17.0	122

PORSCHE 911 AND DERIVATIVES

	Year	List price	Curb weight, lb	Wheelbase, in.	Engine, cc	Bhp @ rpm	0–60 mph, sec	0–1320 ft (¼ mi), sec	Maximum speed, mph
911	1965	$6500	2360	87.0	1991	145 @ 6100	9.0	16.5	132
911S	1967	$7074	2365	87.0	1991	180 @ 6600	8.1	15.7	141
911 Sportomatic	1968	$6150	2410	87.0	1991	148 @ 6100	10.3	17.3	117
911E	1969	$7050	2361	89.3	1991	160 @ 6500	8.4	16.0	130
911S 2.2	1970	$8750	2390	89.3	2195	200 @ 6500	7.3	14.9	144
911T Sportomatic	1971	$6595	2390	89.3	2195	142 @ 5800	9.1	17.2	122
911E 2.4	1972	$9078	2485	89.4	2341	165 @ 6200	6.6	15.4	138
911S 2.4	1973	$10,160	2570	89.4	2341	181 @ 6500	7.8	16.3	142
Carrera RSR	1973	$14,000	1850	89.4	2806	280 @ 8000	5.6	13.2	178
911 Carrera 2.7	1974	$13,575	2490	89.4	2687	167 @ 5800	7.5	15.1	144
911 Carrera	1975	$13,625	2525	89.4	2687	152 @ 5800	8.2	16.5	134
Turbo Carrera	1976	$25,800	2785	89.4	2993	234 @ 5500	6.7	15.2	156
Group 4 Turbo	1977	$40,000	2655	89.4	2993	485 @ 7000	5.8	14.2	163
Group 5 Turbo	1977	$75,000	2340	89.4	2857	590 @ 7900	3.3	8.9	183
911SC	1978	$17,950	2740	89.4	2994	172 @ 5500	6.3	15.3	126
911 Turbo	1978	$34,000	2960	89.4	3299	253 @ 5500	5.0	13.7	156

4-CYLINDER PORSCHES

	Year	List price	Curb weight, lb	Wheelbase, in.	Engine, cc	Bhp @ rpm	0–60 mph, sec	0–1320 ft (¼ mi), sec	Maximum speed, mph
356-4 Coupe	1952	$4208	1850	82.7	1488	65 @ 4800	13.8	18.4	103
Super Coupe	1954	$4395	1860	82.7	1488	70 @ 5000	12.4	18.4	108
1500S Speedster	1955	$3495	1790	82.7	1488	84 @ 5000	10.3	17.3	101
Continental Coupe/Speedster	1956	$3590/$2995	1920/1750	82.7	1488	66 @ 4400	15.0/13.9	19.8/19.2	98/95
Carrera Coupe	1956	$5995	2035	82.7	1498	110 @ 6200	11.5	17.7	120
550 Spyder	1957	$6800	1510	82.7	1498	137 @ 6200	8.2	16.1	122
1600 Coupe	1957	$3790	1930	82.7	1582	70 @ 4500	14.4	19.3	101
1600 Super Speedster	1958	$3928	1790	82.7	1582	88 @ 5000	10.5	17.1	105
1600 Convertible D	1959	$3695	1900	82.7	1582	70 @ 4500	14.0	19.4	98
Super 90	1960	$4195	1935	82.7	1582	102 @ 5500	12.5	17.7	117
1600 Normal Coupe	1961	$4095	1980	82.7	1582	70 @ 4500	14.4	19.4	100
Super 90 Competition	1961	$5551	1780	82.7	1582	102 @ 5500	11.2	16.7	89
Carrera 2-liter	1962	$7595	2220	82.7	1966	152 @ 6200	9.2	16.9	123
356-C	1964	$4295	1970	82.7	1582	88 @ 5200	13.5	18.9	100
904 GTS	1964	$7425	1350	90.6	1966	198 @ 7200	6.4	14.5	150
912	1966	$4700	2100	87.0	1582	102 @ 5800	11.6	18.1	119
914	1970	$3595	2085	96.4	1679	85 @ 4900	13.9	19.2	109
914 2-liter	1973	$5299	2145	96.4	1971	91 @ 4900	10.3	17.8	119
912E	1976	$10,845	2395	89.4	1971	86 @ 4900	11.3	18.2	115
924	1976	$9400	2415	94.5	1984	95 @ 5500	11.9	18.3	111
924	1977	$9995	2575	94.5	1984	110 @ 5750	11.0	18.0	117
924 Automatic	1979	$14,600	2825	94.5	1984	110 @ 5750	12.4	19.3	—
924 Turbo	1979	est $21,000	2835	94.5	1984	143 @ 5500	7.7	16.3	132

ROAD & TRACK
R&T
ROAD TEST
S · 04843

PORSCHE 924 TURBO

Porsche puts in the perFOURmance

BY JOHN DINKEL
Editor, Road & Track

PHOTOS BY WM A. MOTTA

THIS IS AN historic road test in several respects. The U.S. version Porsche 924 Turbo described herein represents the 1000th road test R&T conducted since we published a test of the 1947 Ford V-8 way back in June 1947. This test is also significant because it represents the first time R&T traveled outside the borders of the U.S. to conduct a full road test. It wouldn't have been possible except for the compact nature of our new test equipment. "Have computer, will travel," is the way I refer to this marvel of space-age technology.

That the 924 Turbo was the subject of our 1000th road test is more than happenstance. We knew several months in advance that R&T was approaching that milestone, so the staff met and generated a list of possible candidates worthy of the honor. It took a bit of haggling but finally the list was whittled down from eight cars to three cars to finally only one.

The 924 Turbo is really the ideal choice. There's hardly a marque that has figured more prominently in the pages of R&T these past 33 years than Porsche. Forty-six of the 1000 road tests R&T has conducted have involved Porsches, starting with the test of the 356-4 in November 1952. It's hard to imagine two cars as different as the 356 and the 924. Yet there are common bonds between these models: Each is Volkswagen-based and carries the Porsche nameplate. And just as that humble 356 evolved into such potent performers as the Carrera and the 550 Spyder, the 924 in Turbo trim has shed its Milquetoast image. Porsche will continue to sell the 911 for as long as it can be modified to meet U.S. safety and emissions requirements and as long as there is demand for the model. But it's obvious that the 924 is THE PORSCHE of the immediate future, and that this is the model the factory is concentrating its development and racing efforts on.

Choosing the 924 Turbo as subject of R&T's 1000th road test was simplicity itself, compared to the logistics involved in getting an early test of the new model. At the time the U.S. version was nowhere near being available for testing, much less for sale, in North America, and yet there seemed little point in journeying to West Germany to test the higher-powered European version that was already on sale there. What we wanted was a full road test of the model American customers would be able to buy, and we were willing to travel if the Porsche people could provide the car.

When we explained our plight to the Porsche public relations man he telexed Zuffenhausen and the factory agreed to provide R&T with one of the very first 924 Turbos built to U.S. specifications. But it would not be simply a matter of traveling to the east coast of the U.S., as we often do for early tests, but to Germany.

This we did, even though the weather there in the middle of winter is hardly conducive to road testing. Thankfully, the sun came out long enough for Art Director Bill Motta to get his pictures and the snow and ice melted (most of it anyway), allowing me to get some important data. All the performance testing was conducted at Porsche's Weissach Proving Grounds and I am indebted to Rolf Hannes for closing the track to Porsche testing for three hours, which allowed me to make my acceleration runs in two directions, canceling the effects of wind, elevation changes and a few nasty ice patches.

You won't find panic stopping distances listed in the data panel for the 4-wheel disc brakes Porsche has developed for the Turbo. Neither will you find skidpad or slalom performance. The day I tested, the Weissach straightaway was slightly damp, not enough to affect acceleration times but enough to lengthen stopping distances and slalom times. I had looked forward to testing on the Weissach skidpad, made famous by Mark Donohue's testing of the 917-30 Turbo, but patches of ice dotted the area and put serious testing on the skids. And though you might think that top speed would be easy to measure on one of Germany's speed-limit-free *Autobahnen*, the day we chose to conduct this test the weatherman presented us with dense fog and blinding snow. This test, as it was to turn out, would have to be carried out in a much less likely place: speed-limited America.

Constant readers of R&T road tests will have noticed a certain difference between this test and almost every other one we publish: a byline. There's a good reason why R&T road tests are usually anonymous. A road test is a staff effort with everyone giving his or her input. In effect, the writer is putting down on paper the collective judgments of the staff, not his personal opinions. We obviously couldn't ship the entire staff to Germany to drive the U.S. 924 Turbo, so the opinions expressed here are mine and Bill Motta's. If we've been less than objective in any of our comments or judgments, and I don't think we have, we hope this explanation of the special circumstances will compensate.

Now to the car. I flew to Germany with an open mind but, frankly, I was mildly pessimistic about the Turbo. R&T's feelings about the anemic performance of the normally aspirated 924 are well known, and I wasn't sure that even the collective wizardry of the Zuffenhausen braintrust could work the needed miracle: transforming the noisy, gutless Audi 4-banger into an engine worthy of the Porsche name. Not to worry. With a quarter-mile time of 16.3 seconds and 0–60 mph in just 7.7 sec, the 924 Turbo isn't going to have rubber dust kicked into its windshield by many sports cars. In fact, except for a few big-engine American cars such as the Corvette and exotics such as the Lamborghini Countach and Ferrari Boxer, the 924 Turbo's main competition is going to come from sibling rivals, the 911SC and the 928.

This significant improvement in performance (the stock 924 does 0–60 in 11.0 sec and covers the quarter mile in 18.0 sec) would probably satisfy most Porschephiles, but I expect more from an engine than sparkling straight-line acceleration—easy starting, for example. The U.S. Turbo is equipped with a 3-way catalyst and in temperatures ranging from 10 degrees below ⟫

AT A GLANCE

	Porsche 924 Turbo	Datsun 280ZX	Ferrari 308 GT4
List price	$20,875	$9899	$38,460
Curb weight, lb	2835	2900	3405
Engine	inline 4	inline 6	V-8
Transmission	5-sp M	3-sp A	5-sp M
0-60 mph, sec	7.7	10.2	7.8
Standing ¼ mi, sec	16.3	18.1	16.0
Speed at end of ¼ mi, mph	88.0	80.0	89.0
Stopping distance from 60 mph, ft	na	160	167
Interior noise at 50 mph, dBA	70	69	77
Lateral acceleration, g	na	0.760	0.779
Slalom speed, mph	na	56.5	57.9
Fuel economy, mpg	est 21.0	21.0	14.0

freezing to around 50 degrees Fahrenheit, the engine fired right up with just a few kicks from the starter. And whether hot, cold or somewhere in between, it exhibited excellent warmup and drive-ability characteristics.

Of critical importance for any turbocharged engine is throttle response. If the engine runs like gangbusters above 4500 rpm but coughs, sputters and lacks tractability at the low end, the car won't be any fun at all in the normal cut and thrust of U.S. driving. Fear not, flexibility as well as quickness are this engine's strengths. Boost starts to build at a low 1600 rpm and maximum boost of 7 psi is achieved at 2800 revs for the smaller but quicker revving turbo used on the U.S. engine. (Comparable figures for the European Turbo are 1800 rpm and 10-psi boost at 2800 rpm.) A sensitive driver will be able to detect the turbo starting to spin at that 1600-rpm figure and by 3000 the engine is twisting at maximum torque. Such a design gives the Turbo the low-end as well as the top-end performance lacking in the normal 924.

Transmission gearing is identical for the U.S. and European models except for 5th, which is a 0.60:1 ratio compared to 0.71:1 for the European model. But the U.S. gets a 4.71:1 final drive compared to 3.17:1, so the American Turbo has shorter gearing across the board. If you drive a normal 924 in top gear at 50–60 mph and step on the throttle, you'll wait all day for something to happen. Not so the Turbo. Even though 5th is an over-overdrive (60 mph is only 2280 rpm), you can let the revs drop to below 2000 in top gear, tromp the go pedal and let the engine pull smoothly, albeit slowly. Try that with a stock 924 and you'll be greeted with a chorus of shakes, shudders, buzzes and groans of protest that won't stop until you downshift at least two gears.

The subtlety of the 924 Turbo is quite a contrast to the 930 Carrera that lets you know it's turbocharged with a neck-snapping bang. The 924 Turbo's forgiving nature was especially

appreciated on the snow- and ice-covered roads around Stuttgart. I never feared an "incident" as I might have with the 930's more peaky boost.

Our other major complaint with the normal 924 is engine noise. Porsche engineers have expended a lot of effort to alleviate this problem. For example, there's a significant increase in sound-deadening material and a thick underhood pad. The engine still isn't quiet but the turbo smooths and softens most of the noise and lowers the frequency so that the engine isn't as buzzy or as jangly on the nerves. And the blower's mild muted whistle lets you know there's a turbo engine under the hood without being blatant about it.

Once you get used to the 5-speed pattern, the gearbox and shifting feel about the same as in the normal 924. Even in the standard driveline there's a fair amount of inertia, and with a larger-diameter clutch and driveshaft and heftier gears and shafts there's more in the 924 Turbo. As a result you can't execute the lightning-quick gear changes possible with cars having the

gearbox directly coupled to the engine; rather a smooth, precise, unhurried motion should be adopted to prevent mild gear clash.

You'll find very few changes inside a 924 Turbo. The instrumentation is standard 924 fare; 1979 Turbos had a 160-mph speedometer with the redline area at high noon, but of course current models are afflicted, like everything else, with the NHTSA's 85-mph proscription. Oddly, there's no turbocharger boost gauge; Porsche explains the omission with the 924's smooth turbo boost. Just the same, we'd prefer having a gauge. The Turbo's steering wheel is a 3-spoke design borrowed from the 911SC. It's the same diameter as the regular 924 wheel, so wheel/thigh clash is still a problem for some drivers. Thankfully, a 4-spoke wheel that's about an inch smaller in diameter is now an option. The seats in our test car had cloth inserts in a handsome ⟫

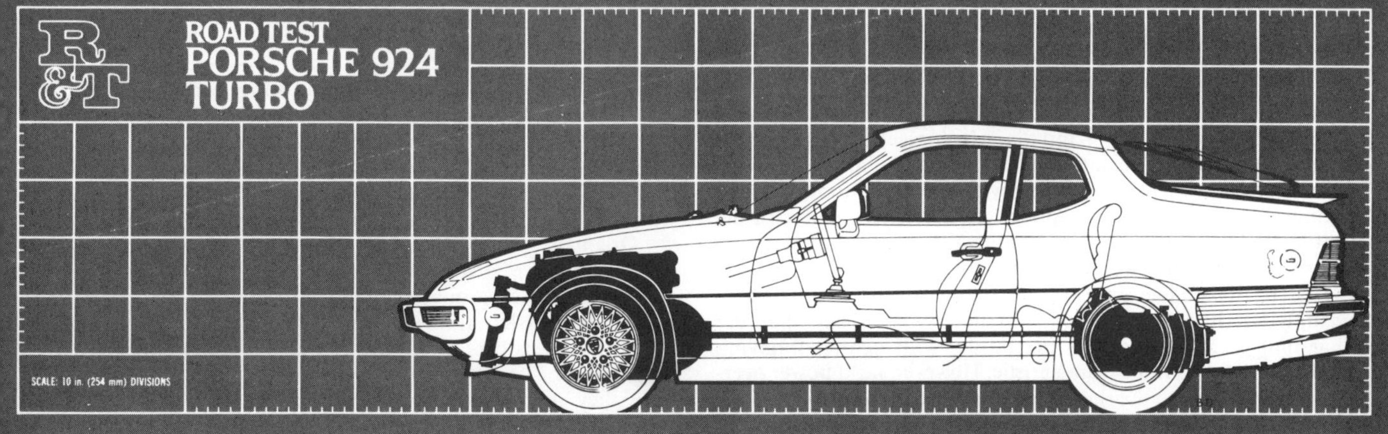

PRICE

List price, all POE $20,875
Price as tested $23,645

Price as tested includes standard equipment (elect. window lifts), rear wiper ($235), AM/FM stereo/cassette ($655), leather-covered steering wheel ($185), Convenience Package, incl air cond, elect. adj heated outside mirrors, removable roof panel and power antenna ($1695)

IMPORTER

Porsche-Audi Div, VW of America, 818 Sylvan Ave, Englewood Cliffs, N.J. 07632

GENERAL

Curb weight, lb/kg	2835	1287
Test weight	2980	1353
Weight dist (with driver), f/r, %		49/51
Wheelbase, in./mm	94.5	2400
Track, front/rear	55.8/54.8	1418/1392
Length	170.1	4321
Width	66.3	1685
Height	50.0	1270
Ground clearance	5.9	150
Overhang, f/r	34.4/41.2	874/1046
Trunk space, cu ft/liters	10.4 + 7.9	295 + 224
Fuel capacity, U.S. gal./liters	16.4	62

INSTRUMENTATION

Instruments: 160-mph speedo, 8000-rpm tach, 99,999 odo, 999.9 trip odo, oil press., coolant temp, fuel level, clock

Warning lights: oil press., brake system, handbrake, battery, low fuel, oxygen sensor, rear-window heat, seatbelts, hazard, high beam, directionals

ENGINE

Type		sohc inline 4
Bore x stroke, in./mm	3.41 x 3.32	86.5 x 84.4
Displacement, cu in./cc	121	1984
Compression ratio		7.5:1
Bhp @ rpm, SAE net/kW	143/107 @ 5500	
Equivalent mph / km/h		147/237
Torque @ rpm, lb-ft/Nm	147/199 @ 3000	
Equivalent mph / km/h		79/127
Fuel injection		Bosch K-Jetronic
Fuel requirement		unleaded 91-oct

Exhaust-emission control equipment: 3-way catalyst

DRIVETRAIN

Transmission	5-sp manual
Gear ratios: 5th (0.60)	2.83:1
4th (0.93)	4.38:1
3rd (1.22)	5.75:1
2nd (1.78)	8.38:1
1st (3.17)	14.93:1
Final drive ratio	4.71:1

ACCOMMODATION

Seating capacity, persons		2 + 2
Head room, f/r, in./mm	37.0/32.0	940/813
Seat width	2 x 19.5/2 x 15.0	2 x 495/2 x 381
Seat back adjustment, deg		50

MAINTENANCE

Service intervals, mi:

Oil/filter change	7500
Chassis lube	none
Tuneup	15,000
Warranty, mo/mi	12/unlimited

CHASSIS & BODY

Layout	front engine/rear drive
Body/frame	unit steel

Brake system....11.1-in. (283-mm) vented discs front, 11.4-in. (289-mm) vented discs rear; vacuum assisted

Swept area, sq in./sq cm	264	1703
Wheels		cast alloy, 15 x 6J
Tires		Pirelli CN36, 185/70VR-15
Steering type		rack & pinion
Overall ratio		22.4:1
Turns, lock-to-lock		4.0
Turning circle, ft/m	33.8	10.3

Front suspension: MacPherson struts, lower A-arms, coil springs, tube shocks, anti-roll bar

Rear suspension: semi-trailing arms, torsion bars, tube shocks, anti-roll bar

CALCULATED DATA

Lb/bhp (test weight)	20.8
Mph/1000 rpm (5th gear)	26.3
Engine revs/mi (60 mph)	2280
Piston travel, ft/mi	1260
R&T steering index	1.35
Brake swept area, sq in./ton	177

RELIABILITY

Owners of earlier-model Porsches reported 11 problem areas and 4 disabling reliability areas compared to overall Owner Survey averages of 12/7. So we expect the overall reliability of the 924 Turbo to be better than average.

ROAD TEST RESULTS

ACCELERATION

Time to distance, sec:

0–100 ft	3.2
0–500 ft	8.7
0–1320 ft (¼ mi)	16.3
Speed at end of ¼ mi, mph	88.0

Time to speed, sec:

0–30 mph	2.4
0–60 mph	7.7
0–90 mph	17.3

SPEEDS IN GEARS

5th gear (4950 rpm)	est 132
4th (6600)	113
3rd (6600)	88
2nd (6600)	62
1st (6600)	35

FUEL ECONOMY

Normal driving, mpg	est 21.0
Cruising range, mi (1-gal. res)	385

HANDLING

Lateral accel, 100-ft radius, g	na
Speed thru 700-ft slalom, mph	na

BRAKES

Minimum stopping distances, ft:

From 60 mph	see text
From 80 mph	see text
Control in panic stop	see text
Pedal effort for 0.5g stop, lb	21

Fade: percent increase in pedal effort to maintain 0.5g deceleration in 6 stops from 60 mph ....nil

Parking: hold 30% grade?	na
Overall brake rating	see text

INTERIOR NOISE

Idle in neutral, dBA	58
Maximum, 1st gear	83
Constant 30 mph	67
50 mph	70
70 mph	73
90 mph	77

SPEEDOMETER ERROR

30 mph indicated is actually	27.0
60 mph	58.0
80 mph	80.0

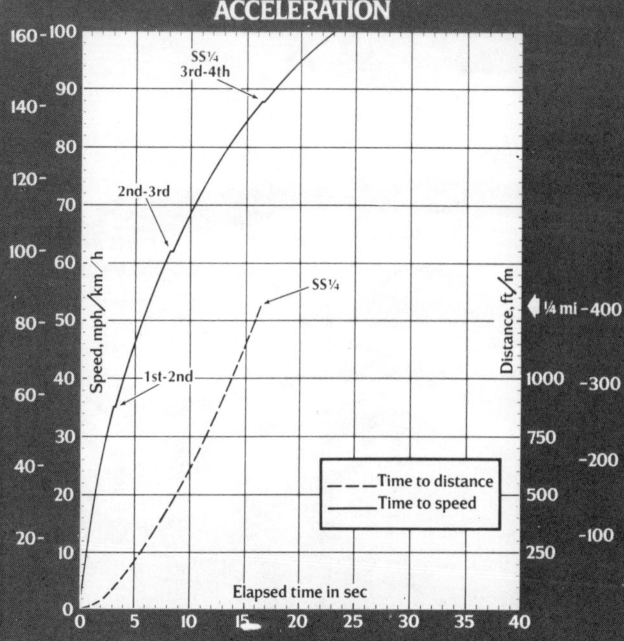

tartan pattern of green, brown, blue, red and white. The same pattern was used on the door panels; the rest of the interior was in accenting shades of brown and tan, which nicely complemented the Teutonic exterior paint scheme of tan and chocolate brown. Our test car was also equipped with air conditioning and a removable roof panel, two options we had little use for in the bitter cold of a German winter but two that can be expected to prove quite popular with American buyers.

Under the circumstances it was a little difficult getting a real handle on handling. In general, though, we'd say that the handling is as balanced as previously, with gentle, forgiving understeer the normal characteristic. However, mild power over-steer is available to the skilled driver and the extra power allows the chassis' real potential to be realized. The ride is firm but certainly not objectionable and the new 4-wheel discs, a combination of 911, 928 and a few new pieces, went about their job unobtrusively, registering 21 lb for a 0.5g stop from 60 mph and exhibiting no fade at the end of our 6-stop sequence.

The only minus in this list of pluses is the steering. It's precise, accurate and affords adequate road feel, but it's slow. Four turns lock-to-lock combined with a 33.8-ft turning circle result in a R&T steering index of 1.35, unusually large for a sports car. This is quite apparent in tight turns, where excessive wheel cranking is required. And the problem is accentuated in the Turbo because the steering ratio was slowed from 19.2:1 to 22.4:1, reflecting a lengthening of the steering arms to improve the front geometry.

For two tankfuls of unleaded fuel, including track testing and

as much hard driving as the ice and snow permitted, the U.S. Turbo returned 17–21 mpg. The EPA city-driving figure is 19 mpg for the 1980 model, and in view of the EPA highway figure of 31 mpg I would expect an overall, everyday mileage of something like 21–22 for most American drivers: a relatively good level of fuel efficiency for a machine of such potent performance.

Originally, the Porsche people thought of offering a 924 Turbo with standard 924 chassis tuning, but they decided not to make this combination available after all. To me this certainly makes sense; Porsche has never compromised the handling and braking potential of its cars to the dictates of a country's speed limits, regardless of how ridiculous they might be. I for one am glad to see they aren't about to start what could set a dangerous precedent. What will be offered besides the full-fledged Turbo are the normal 924 and 100 standard-engine 924s equipped with the Turbo chassis, just what the enthusiasts ordered for SCCA D Production racing.

Another interesting fact: With the revised aerodynamics and the rear spoiler, the coefficient of drag is reduced from 0.36 to 0.35. And Porsche has purposely made the tail spoiler integral with the rear window (and thus very expensive) because install-ing this device on a normal 924 could increase the maximum speed to the point that the engine would exceed redline.

The price goes hand in hand with the Turbo's top speed. It's up also, to $20,875. This is certainly disappointing news for those who expected the Turbo to reach our shore at a more affordable price. You can credit (or debit) inflation and the expense involved in meeting U.S. safety and emission regulations with a low-volume model for a hefty portion of that 5-digit price tag. If this explanation is little consolation, consider the following. When the 930 Turbo Carrera was introduced in 1976 it listed for $26,000. By the time the last one was sold in early 1980, the price had escalated to $42,520. So although the entry fee for a 924 Turbo starts at nearly $21,000, you're not purchasing merely a car, but an investment. And a 924 Turbo is 1000 percent more fun to own than some stodgy old shares of IBM or Ma Bell.

1980 PORSCHE 924

On its way to becoming a true Porsche

PHOTOS BY JOE RUSZ

WELCOME TO YET another chapter in the continuing saga of the Porsche 924, the loss leader of Zuffenhausen's line. You may recall that in our last installment *(R&T's Guide to Sports & GT Cars 1979)*, we called the car "a fine little GT," but one that "falls short of the mark because of a number of niggling problems." The car in question was a 1978 model, an improved version of the 1977½ tested previously and substantially better than the original 924 tested in July 1976. If anything, it's evident that Porsche is cleaning up the car's act, and the 1980 version is proof that the program is succeedin

The latest 924 with its 3-way catalytic converter cum oxygen sensor emission control system is the best of the econo-Porsche lot. The car is quicker, more frugal with fuel and has better driveability than its predecessors. And it is still very much a 924 which is bad and good. Here's what we mean.

From the outset, the 924's major problems have been ride, noise and lack of performance. Of ride, we've said it's "jouncy and harsh" with little compliance; of the engine, "buzzy" and "not very quick for a car of its nature." That's the bad news; here's the good. For 1980, Porsche has made significant improvements in all these areas so the latest version is to its predecessors what the stereo receiver/amplifier is to the superheterodyne.

Let's take ride. The first 924 had a bad case of freeway hop that thoroughly annoyed our staff and prompted two of our staff members to swear that they "would refuse to buy the (1976) 924

for that reason alone." We're not sure that our comments had anything to do with it, but at this point it seemed that Porsche engineers rolled up their sleeves and went to work, tightening installation tolerances and, later, adding rubber rear suspension mountings and hydraulic transmission mounts. The dreaded freeway hop was subdued, although compliance, the suspension's ⇒

<table>
<tr><td colspan="4">AT A GLANCE</td></tr>
<tr><td></td><td>Porsche
924</td><td>Datsun
280ZX</td><td>Mazda
RX-7 GS</td></tr>
<tr><td>List price</td><td>$16,770</td><td>$9899</td><td>$8295</td></tr>
<tr><td>Curb weight, lb</td><td>2680</td><td>2900</td><td>2420</td></tr>
<tr><td>Engine</td><td>inline 4</td><td>inline 6</td><td>Wankel</td></tr>
<tr><td>Transmission</td><td>5-sp M</td><td>3-sp A</td><td>5-sp M</td></tr>
<tr><td>0–60 mph, sec</td><td>10.6</td><td>10.2</td><td>9.2</td></tr>
<tr><td>Standing ¼ mi, sec</td><td>17.6</td><td>18.1</td><td>17.0</td></tr>
<tr><td>Speed at end of ¼ mi, mph</td><td>78.0</td><td>80.0</td><td>83.0</td></tr>
<tr><td>Stopping distance from 60 mph, ft</td><td>150</td><td>160</td><td>151</td></tr>
<tr><td>Interior noise at 50 mph, dBA</td><td>70</td><td>69</td><td>68</td></tr>
<tr><td>Lateral acceleration, g</td><td>0.766</td><td>0.760</td><td>0.779</td></tr>
<tr><td>Slalom speed, mph</td><td>58.9</td><td>56.5</td><td>59.3</td></tr>
<tr><td>Fuel economy, mpg</td><td>23.0</td><td>est 21.0</td><td>22.0</td></tr>
</table>

ability to absorb small, sharp bumps, still left something to be desired—until the 1980 model came along. Suddenly, the tone of our testers comments changed: "The ride is much improved," said one. "It's firm as always, but much of the terrible harshness and vibration of earlier versions has been isolated and/or damped to a livable level. It's certainly not perfect, but it is sufficiently better to rate with the best in its class."

Addressing ourselves to the problem of engine and driveline/suspension noise, we found that, once again, the 1980 model offers dramatic proof that Porsche is building a better 924. Why didn't they get it right in the first place, you ask? Perhaps because *Rom wurde nicht in einem Tag gebaut.*

By nature many inline 4-cylinders are rough runners and Porsche's sohc 1984-cc powerplant is no exception. We've complained about engine resonance in the past and said things like "noisy, noisy, noisy." If you expected us to say we didn't notice it in the 1980 model, you're wrong. But the racket is a bit less objectionable because Porsche has done a pretty decent job of masking it with a new sound-deadening package. As a matter of fact, the improvements in sound deadening are also evident in the body, the 924's other area of objectionable noise. Although there have never been any rattles and the like associated with this car, its tight body structure has acted like a drum, amplifying road noise transmitted by the suspension and driveline. In the past this resulted in booming when the car was driven over coarse road surfaces. But again, Porsche has tackled this problem and our 1980 test car shows it: "There's much less body thumping

We're not about to yell ourselves hoarse extolling the merits of the 1980-spec engine. But, by God, it is markedly better, thanks mainly to the use of a 3-way catalytic converter emission control system. More precise control of combustion/emissions has enabled Porsche to raise the compression ratio to 9.0:1 and to increase horsepower to 115 bhp. As a result, 0 to 60 and quarter-mile times have both dropped by 0.4 sec and the engine pulls to an indicated 6600 rpm in 4th and 4900 rpm in 5th, well over the century mark. Now the 924 feels like a sports car rather than a stylish stone. Even more important than this wide-open-throttle performance is the improvement in low- and mid-range-throttle response, previously a 924 weak point. Cold-start driveability is excellent and (here's the best news) fuel consumption has been reduced by almost 20 percent! With gasoline prices around $1.20 per gallon, such news *is* worth shouting about.

In their quest for better fuel economy, Porsche engineers wisely chose to maintain last year's overall drive ratio of 3.00:1 in top gear. But they also hoped to improve low-end acceleration and for this, shorter gearing was in order. So the factory compromised by lowering the gearbox ratios and raising the ring and pinion ratio. Thus, the old 5-speed's 2.79, 1.68, 1.11, 0.81 and 0.60:1 1st through 5th gear ratios and 5.00:1 final drive gave way to the 1980 model's 3.60, 2.13, 1.36, 0.97 and 0.73 gearing and 4.11:1 final drive.

In addition to changing the gearing, Porsche revised the shift pattern so that reverse is at the far right and down, just below 5th. In the 1979 model, reverse was to the left and up, just out of the H

over lane dots and what noise is transmitted results from the firm suspension and the high performance tires, not from the booming of the body cavity," said one editor who summed up his comments by calling the latest 924 a "much more refined and quiet car."

This brings us to the 924's last major problem area, performance. Our feeling is that it has never been commensurate with the car's handling, which is excellent. Through the years, Porsche has strived to improve acceleration and driveability, raising the compression ratio from 8.0:1 to 8.5:1 and the horsepower from 95 to 110 bhp. Changes in camshaft and ignition timing and the use of a lower (higher numerical) rear axle ratio (3.88 vs 3.44:1) resulted in improved low- and mid-range acceleration in the 1977½ model which, like the 1978 version, did 0 to 60 in 11.0 seconds and turned the quarter mile in 18 sec flat. A bit better perhaps, but as we said at the time: "Not very quick for a car of its nature." And let's not forget that the driveability was not great. Nor was fuel economy—18.5 mpg was nothing to shout about.

from 1st and many a time we'd find ourselves accidentally engaging reverse while hunting for low at a traffic light.

From its inception, the 924 has had one shortcoming that's annoyed many drivers. The car's steering column is set too low and as a result, there is interference between the wheel and the driver's thighs. With some drivers (you know who you are) this interference occurs even while the wheel is straight, but with others it occurs when the eccentrically positioned wheel is turned 180 degrees. Then hands collide with thighs and spirited driving suffers. We've complained about this in our road tests and to Porsche personnel who finally offered a quick and easy solution: For reasons unclear to us (seat/floor relationship, perhaps), the seats are mounted on ½-in. aluminum spacers. We removed these, and the extra clearance makes all the difference.

More than almost any other production GT, the 924 has the handling, braking and the performance (almost) that one expects from a proper sports car. It's extremely well balanced with neutral handling that verges on understeer with the power on and oversteer when the power is suddenly turned off. "Backing off the throttle at speed unloads the car slightly and can result in mild oversteer that can catch the unwary," cautioned the Editor. But he went on to praise the 924's high-speed stability, especially in our slalom test. And he summarized all of our feelings saying, "This car makes twisty roads a driver's delight."

In keeping with the Porsche tradition are the 924's brakes—discs up front and drums at the rear. Unchanged from 1979, they exhibit no locking or fade and are capable of stopping the car in a straight line every time with very good control. Want even more stopping power? Then order the optional 4-wheel discs.

The 924's interior which we once called, "functional, austere, Teutonic," is the same though the instruments have new faces. This means that the speedometer now peaks at 85 mph in accordance with federal regulations. That's ludicrous in a car designed to travel safely in excess of such speeds.

Also unchanged and thus unimproved is the 924's ventilation system. The controls are a bit vague and cool airflow is minimal. In fact, the only way to get sufficient cool air inside the cabin is to engage the air conditioning or to open (by removing) the sunroof.

One controversial area of the 924's interior are the seats. Some staffers found them hard and unyielding while others appreciated their firmness and exceptional support. We'll get to the bottom of this problem, but in the meantime, we'd like to compliment Porsche for the new car's quietness. At cruising speeds of 50 to 70 mph, the 1980 924 is approximately five dBA quieter than the 4-speed model we last tested.

Unfortunately, inflation and the disparity between the dollar and the Deutsche Mark are gaining on the 924 and the car that once cost $10,000 is now an astounding $15,970 base. Add the price of an AM/FM/stereo radio ($200—$400) and perhaps the handling package ($1960, which includes 4-wheel-disc brakes, a larger-diameter front anti-roll bar, a rear bar, a leather steering wheel, sport shocks and alloy wheels with Pirelli P7 tires) and you have a car that will easily cost $20,000 after all requisite fees are paid. Contrast this with the Datsun 280ZX ($9899, base) or Mazda RX-7 GS ($8295, base) and you'll see why the German GT is taking such a beating in the market place.

Still, the 924 is very much a car in the Porsche tradition, even though it's built in Neckarsulm (by Audi) and not in Zuffenhausen (by Porsche). The 1980 model comes even closer to being a thoroughbred, especially where styling, quality, handling and, yes, even performance are concerned. And let's not forget that 23.0-mpg fuel consumption figure. Perhaps one eloquent staff member summed it up best when he said, "A day in the mountains in this car is a sheer delight. It's almost enough to make you forget the price."

PRICE	
List price, all POE	$16,770
Price as tested	$19,055
Price as tested includes AM/FM stereo cassette ($665), E82 Package: air cond, removable sunroof, elect. adj heated outside mirrors, leather-wrapped steering wheel ($1620)	

GENERAL

Curb weight, lb/kg	2680	1217
Test weight	2890	1312
Weight dist (with driver), f/r, %		48/52
Wheelbase, in./mm	94.5	2400
Track, front/rear	55.9/54.0	1420/1372
Length	170.1	4321
Width	66.3	1684
Height	50.0	1270
Trunk space, cu ft/liters	10.4 + 7.9	295 + 224
Fuel capacity, U.S. gal./liters	16.4	62

ENGINE

Type	sohc inline 4
Bore x stroke, in./mm	3.41 x 3.32 ....86.5 x 84.4
Displacement, cu in./cc	121 1984
Compression ratio	9.0:1
Bhp @ rpm, SAE net/kW	115/86 @ 5750
Torque @ rpm, lb-ft/Nm	111/151 @ 3500
Fuel injection	Bosch K-Jetronic
Fuel requirement	unleaded, 91-oct

DRIVETRAIN

Transmission	5-sp manual
Gear ratios: 5th (0.73)	3.00:1
4th (0.97)	3.99:1
3rd (1.36)	5.59:1
2nd (2.13)	8.75:1
1st (3.60)	14.80:1
Final drive ratio	4.11:1

CHASSIS & BODY

Layout	front engine/rear drive
Body/frame	unit steel
Brake system	10.1-in. (257-mm) discs front, 9.1 x 1.5-in. (231 x 38-mm) drums rear; vacuum assisted
Wheels	cast alloy, 14 x 6J
Tires	Fulda, 185/70HR-14
Steering type	rack & pinion
Turns, lock-to-lock	4.0
Suspension, front/rear: MacPherson struts, lower A-arms, coil springs, tube shocks, anti-roll bar/semi-trailing arms, torsion bars, tube shocks, anti-roll bar	

CALCULATED DATA

Lb/bhp (test weight)	25.1
Mph/1000 rpm (5th gear)	25.0
Engine revs/mi (60 mph)	2400
R&T steering index	1.23
Brake swept area, sq in./ton	186

ROAD TEST RESULTS

ACCELERATION

Time to distance, sec:

0-100 ft	3.5
0-500 ft	9.4
0-1320 ft (¼ mi)	17.6
Speed at end of ¼ mi, mph	78.0

Time to speed, sec:

0-30 mph	3.3
0-50 mph	7.4
0-60 mph	10.6
0-70 mph	14.0
0-80 mph	18.5
0-90 mph	25.0

SPEEDS IN GEARS

5th gear (4900 rpm)	123
4th (6600)	123
3rd (6600)	84
2nd (6600)	54
1st (6600)	32

FUEL ECONOMY

Normal driving, mpg	23.0

BRAKES

Minimum stopping distances, ft:

From 60 mph	150
From 80 mph	280
Control in panic stop	very good
Pedal effort for 0.5g stop, lb	25
Fade: percent increase in pedal effort to maintain 0.5g deceleration in 6 stops from 60 mph	nil
Overall brake rating	very good

HANDLING

Lateral accel, 100-ft radius, g	0.766
Speed thru 700-ft slalom, mph	58.9

INTERIOR NOISE

Constant 30 mph, dBA	67
50 mph	70
70 mph	76

SPEEDOMETER ERROR

30 mph indicated is actually	30.0
60 mph	60.5

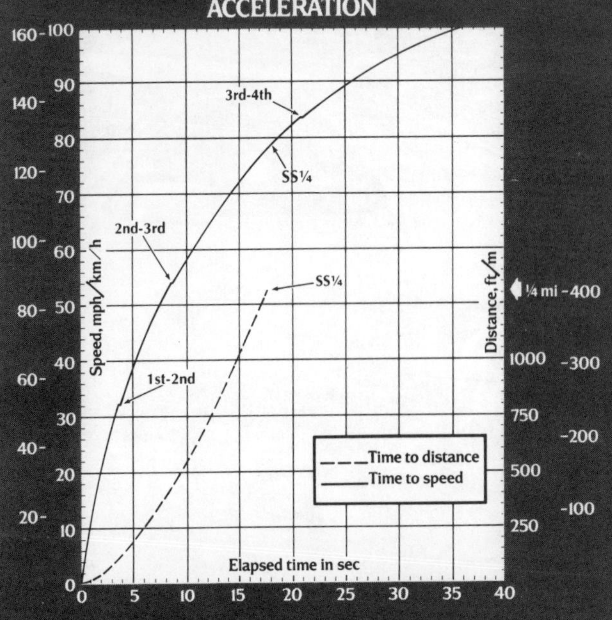

"WE'RE GEARING UP for hard times," said Porsche's racing and public relations director Manfred Jantke at last September's Frankfurt Automobile Exhibition. Jantke knew whereof he spoke: Porsche was heading toward the end of a year that saw business in its main market, the U.S., shift into reverse gear. And the trends of recent times—sharp increases in fuel prices and the weakening dollar—showed every sign of continuing to cut into Porsche's potential for sales.

The main problem for Porsche, however, is not the top-of-the-line 928; it's the 924, whose sales dropped 13 percent from the previous year's level in the period leading up to that Frankfurt show. The 924 is Porsche's volume model, and at $16,000-plus it has a hard time competing with the Mazda RX-7 at less than $10,000. In the 928's stratospheric market segment, however, neither price nor concerns about fuel cost and availability seem to have much effect. A car like the 928 is pretty much above it all. Porsche 928s are no longer commanding the big over-list price pads they did a year or so ago, but with 1555 of them sold in 1979 compared to 1535 during 1978, they're obviously holding up well enough in the American market.

First introduced in Europe at the beginning of 1977, the 928 is what Porsche planners deemed the right formula for a luxury sports-GT car of the Eighties and beyond. It was a big departure—and a huge investment in the future—for the small Stuttgart, West Germany maker of thoroughbred sports cars: Porsche was abandoning its traditional, tried-and-true formula of rear-engine cars with air cooling and relatively small displacement for a front-engine car with a watercooled V-8 of a hefty 4½ liters.

The 928's development had started well before the first big oil shock in 1973, and the move to big-engine technology had been spurred not by energy-conservation considerations but the then predominant priority of emission control: Porsche spokesmen had made it clear some time before the new car's introduction that small-displacement, aircooled engines were tougher to de-smog than big, watercooled ones. Had the 928's gestation period begun in the late Seventies, Porsche might have taken a different route altogether in its conception and development.

Be that as it may, Porsche's flagship is a relatively big, heavy and thirsty automobile. Going into 1980 it remains essentially unchanged, though since its 1977 introduction it has undergone numerous detail changes and improvements. A unit steel body-chassis with extensive aluminum panelwork is the platform, powered by a 4474-cc aluminum V-8 in the so-called "front mid-engine" position and driving the rear wheels through a rear-mounted transaxle. The 928's suspension is all-independent, as it has always been with Porsches, and the brakes are vented discs all around. Porsche fits the 928 with a choice of high-performance Pirelli tires and a list of comfort, convenience and ergonomic features unexcelled by any other maker.

So the 928 is what you might call the Ultimate Sports Car. There are GTs and sports cars that, at least on the surface, seem to be more exotic—see elsewhere in this *Guide*. But in reality, it's the Porsche that has everything and does everything that counts.

For 1980 there are some fairly sweeping changes under the 928's long, low hood, mainly intended to improve the way the big

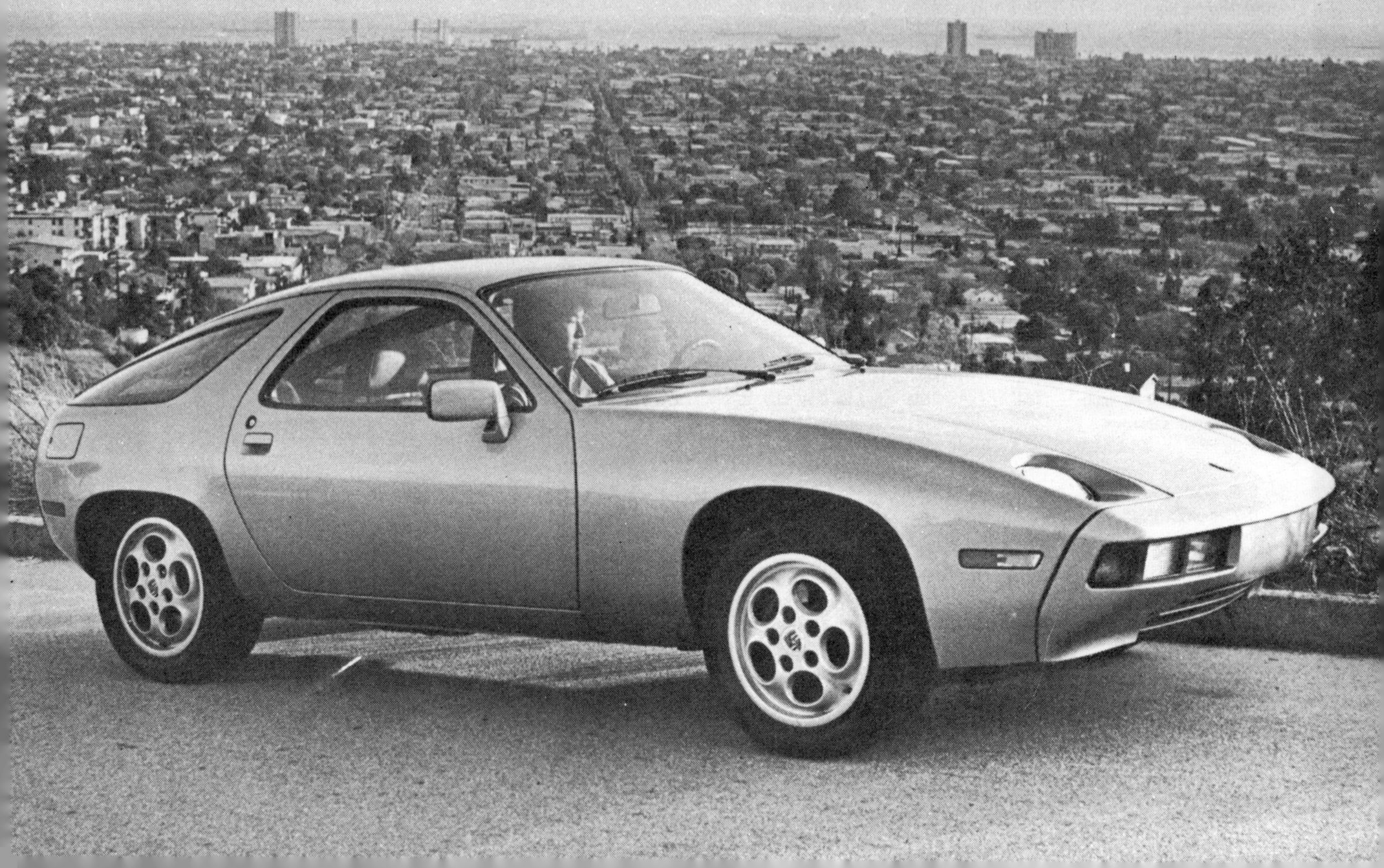

V-8 uses fuel. The engine's compression ratio has been boosted half a point to 9.0:1, its mechanical Bosch K-Jetronic fuel injection replaced by the same firm's electronic L-Jetronic system, and its emission control system completely redone. Gone is the now old-fashioned 2-way catalyst with its attendant exhaust-gas recirculation and air injection, replaced by the all-encompassing combination of a 3-way catalyst and an oxygen sensor. Together with a revision of the ignition timing, these changes result in both better performance and significantly improved fuel economy.

To quantify: The V-8's power is up an insignificant 1 bhp over last year's 219, and its peaking speed has actually moved upward from 5250 to 5500 rpm, which doesn't sound so good. But the torque curve has been fattened by a noticeable amount—the peak is up from 245 lb-ft at 3600 rpm to 265 at 4000—and Porsche claims a 0.4-second improvement in the 0–60 mph time for the 5-speed manually shifted 928 to show for it.

More important in our age of fuel problems, however, is the fuel-economy side. Here the 928 needed improving, and if the EPA figures are indicative of a real gain, then Porsche engineers extracted the expected benefits from the 3-way/sensor emission system. The 5-speed 928's EPA ratings are up from 12 mpg city and 19 mpg highway to 15 and 24; for the automatic version the improvement is from 12/18 to 15/23. Not that all this makes the 928 economical of fuel—but at least it is a substantial gain.

It was the automatic-transmission 928 we tested for the *Guide*. Available as a no-extra-cost alternative to the excellent standard 5-speed manual gearbox, the Mercedes-built automatic is essentially that used in the other Stuttgart carmaker's most powerful V-8 model (the 6.9) but is especially tailored for Porsche requirements and installed in Porsche's own housing to fit into the 928's transaxle.

Mercedes' automatic is one of the best, so the choice was a good one, though not all the changes Porsche incorporated seem the best way to go. One is the selector lever. Mercedes uses a superb gated shifter that allows almost stickshift-style driving for those so inclined, but Porsche chose a T-handle lever with a pushbutton release for selecting the driving ranges other than those most frequently used.

Still, Porsche got good results. The driver of a 928 automatic has complete freedom to downshift from Drive to 2nd without worrying about a detent or lockout, and when this is done the transmission responds virtually instantly. Getting into 1st gear, labeled L, does require a push of the release button, but this is probably best for the safety of the car and its transmission.

Appropriately, the Porsche calibration of this unit is sportier than Mercedes'. Upshifts come relatively late in normal driving; one often has the feeling that the box should go ahead and shift, but it doesn't—but then this means that the 928 remains ready to leap ahead, rather than having to shift back when the driver demands power at lower driving speeds. Likewise the torque converter, which would be described by an automotive engineer as a "loose" one. Moving off from rest, the 928 seems to require too many engine revs and sounds as if it were struggling; but in reality it's doing anything but that. Indeed, the high rotational speed helps keep the engine ready to respond, just like the shift calibration.

Another facet of this automatic's sportiness is that it is also set to upshift at high engine speeds when the accelerator foot is held to the floor, in marked contrast to most self-shifting units. Left in Drive, it takes the engine all the way to its 6000-rpm redline in 1st gear and to its power peak in 3rd; hence there's no need to experiment with manual shifting, as we often do in our acceleration tests in an effort to extract the last bit of through-the-gears performance from automatic transmissions.

Like Porsche's claim for the 5-speed version, the automatic 928 for 1980 also shows proof of its improved torque curve in the performance figures. The test car, for example, reached 60 mph from a standing start in 8.1 sec (our 1978 test of the automatic 928: 8.3 sec) and covered the standing-start quarter-mile in 16.2 sec (16.6). But surprisingly, the 1980 car was doing 2 mph less at the end of the quarter.

As for economy, we had the car only long enough to use one tank of fuel, and that one came in at under 10 mpg for a lot of cold starts and city driving. However, with the 1978 car we measured 15 mpg over a longer period, 3 mpg more than that year's EPA city figure, and giving the 1980 car equal credit we predict something like 18 as a typical everyday figure for today's automatic 928. The manual-shift version should do marginally better.

Other changes for 1980 are mainly in equipment. Leather upholstery, for instance, is now standard, but some other items formerly standard are now optional. All 928s, 5-speed or automatic, now come with 15-in. wheels and Pirelli P6 tires (215/60VR-15) instead of the full-bore 16s and P7s that used to be standard on manually shifted cars; our test car had the P7s, however. The radio-cassette unit has been deleted from the standard list too; now four speakers and a power antenna are installed in all 928s but the radio left out because, says a Porsche spokesman, most customers prefer to choose their own radio-tape units. Our car had the Japanese-made Blaupunkt 2001 system, a $435 item including dealer installation and making a good impression with its decent FM reception and Dolby noise suppression on the cassette side.

Other items now optional include a rear-window wiper and the headlight-washing system; the test car had neither of these, but it ⟫→

AT A GLANCE

	Porsche 928	Maserati Merak/SS	Mercedes-Benz 450SL
List price	$37,930	$34,750	$36,130
Curb weight, lb	3370	3185	3615
Engine	V-8	V-6	V-8
Transmission	3-sp A	5-sp M	3-sp A
0–60 mph, sec	8.1	9.1	11.7
Standing ¼ mi, sec	16.2	17.0	18.6
Speed at end of ¼ mi, mph	89.5	84.5	77.5
Stopping distance from 60 mph, ft	138	151	140
Interior noise at 50 mph, dBA	72	74	69
Lateral acceleration, g	0.811	0.824	0.700
Slalom speed, mph	59.7	52.4	54.2
Fuel economy, mpg	est 18.0	14.0	16.0

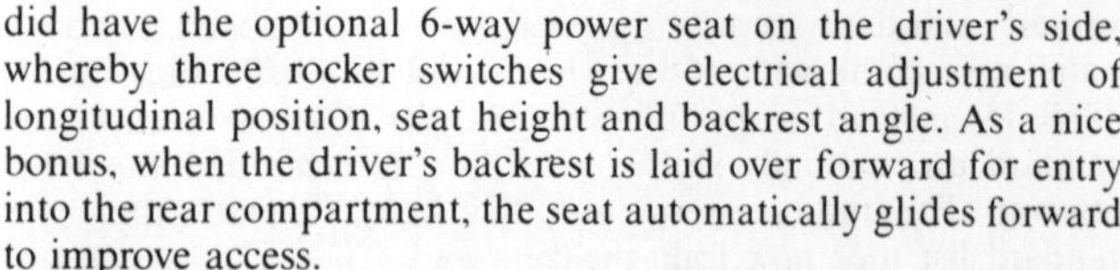

did have the optional 6-way power seat on the driver's side, whereby three rocker switches give electrical adjustment of longitudinal position, seat height and backrest angle. As a nice bonus, when the driver's backrest is laid over forward for entry into the rear compartment, the seat automatically glides forward to improve access.

Such thoughtful details typify the 928's cockpit; it is a veritable treasure box of new, innovative and generally successful ergonomic ideas. In part because of these, it is the most comfortable Porsche ever built, though some drivers object to the feeling of being buried within a deep, spacious cavity and the attendant difficulty of seeing out of it; judging where the car's corners might be for maneuvering in tight places is sometimes very difficult. The instrument pod, containing not only all the primary instruments but a comprehensive array of warning signals that appear as words on what seem to be blank instrument dials, moves up and down with the adjustable steering column so that it is never obscured by the steering-wheel rim.

Electric adjustment for the outside rearview mirror(s) is another such detail. On the left door is a slick hemispherical 4-way switch for adjusting that door's mirror; if the optional right-hand one is ordered, as it was on the test car, an additional (rocker) switch sends the signals to that side's mirror. And the location of the front-rear fader control for the radio system seems just right: on the outboard side of the driver's seat. Not to mention door armrests that can be adjusted some 2 in. farther inboard at the flick of a release lever, or a drop-down vanity mirror flanked with strip lights in the best dressing-room tradition.

Generally fulfilling the best Porsche tradition are the rich leather and fine carpeting of the present 928 interior, as well as most other materials and fits; the feeling is one of great luxury and quality. But there were some slips in the test car. One was that the guide insert where the right-hand seatbelt feeds out of the B-pillar fell out repeatedly, another that the handle for that armrest release lever came off in a passenger's hand, still another that an electric motor outboard of the right front seat was covered by a carpeted box that was only halfway secured. A design, rather than quality, flaw manifests itself when a side window is lowered: The outside mirror on that side generates a highly audible wind whistle, even at city speeds. New since we last tested the 928 is an electric, replacing a vacuum, central locking system. This does its job well enough, though it takes longer than the Mercedes or BMW vacuum systems and makes a noise fascinatingly like that of the landing flaps on a 747.

The 928's comprehensive heating/ventilation/air-conditioning system works very well generally, although there is one drawback to its operation in the ventilation mode. The two middle dash air outlets deliver air only on the air-conditioning setting, and most of our drivers found the adjustable vents in the doors inadequate. For most 928 owners this is probably an academic point, as our experience indicates that they'll use the air conditioning instead of trying to save fuel by using ventilation only, but to us this is a significant failing.

Another possibly academic point was the one about access to the rear seats. It's conceivable that some 928 owners will value the stylish but barely usable +2 rear seating area for their children, pets or the occasional short-term rider, but it cannot be considered a place for grown folks. Just the same, Porsche has taken it very seriously, providing grab handles, an ashtray and lighter, a shallow storage compartment, a map light and sun visors that rotate to shade the leading edge of the big rear ⟫⟫⟫

ROAD TEST

PORSCHE 928

PRICE

List price, all POE		$37,930
Price as tested		$41,160

Price as tested includes standard equipment (air cond, power strg, leather upholstery), 16-in. wheels & Pirelli P7 tires ($1525), auto trans (no chg), elect. adj. driver's seat ($600), elect. right-hand mirror ($110), AM/FM stereo/cassette ($435), dealer prep (est $250), inland freight ($210)

IMPORTER

Porsche-Audi Div, VW of America, 818 Sylvan Ave, Englewood Cliffs, N.J. 07632

GENERAL

Curb weight, lb/kg	3370	1528
Test weight	3550	1610
Weight dist (with driver), f/r, %		50/50
Wheelbase, in./mm	98.3	2497
Track, front/rear	60.8/59.6	1545/1515
Length	175.7	4463
Width	72.3	1836
Height	51.6	1311
Ground clearance	4.7	119
Overhang, f/r	39.7/37.7	1008/958
Trunk space, cu ft/liters	6.3 + 14.2	178 + 402
Fuel capacity, U.S. gal./liters	22.4	85

ACCOMMODATION

Seating capacity, persons		2 + 2
Head room, f/r, in./mm	36.5/927	32.0/813
Seat width, f/r	2 x 20.0/508	2 x 15.0/381
Seat back adjustment, deg		70

ENGINE

Type		sohc V-8
Bore x stroke, in./mm	3.74 x 3.11	95.0 x 78.9
Displacement, cu in./cc	273	4474
Compression ratio		9.0:1
Bhp @ rpm, SAE net/kW	220/164	@ 5500
Equivalent mph / km/h		144/233
Torque @ rpm, lb-ft/Nm	265/359	@ 4000
Equivalent mph / km/h		104/169
Fuel injection		Bosch L-Jetronic
Fuel requirement		unleaded, 91-oct

Exhaust-emission control equipment: 3-way catalyst, oxygen sensor

DRIVETRAIN

Transmission ....automatic; torque converter with 3-sp planetary gearbox

Gear ratios: 3rd (1.00)		2.75:1
2nd (1.46)		4.02:1
1st (2.31)		6.35:1
1st (2.0 x 2.31)		12.70:1
Final drive ratio		2.75:1

MAINTENANCE

Service intervals, mi:	
Oil/filter change	15,000/30,000
Chassis lube	none
Tuneup	30,000
Warranty, mo/mi	12/unlimited

CALCULATED DATA

Lb/bhp (test weight)	16.1
Mph/1000 rpm (3rd gear)	26.0
Engine revs/mi (60 mph)	2310
Piston travel, ft/mi	1197
R&T steering index	0.98
Brake swept area, sq in./ton	251

CHASSIS & BODY

Layout		front engine/rear drive

Body/frame ....unit steel with aluminum doors, hood, front fenders

Brake system ....vented discs; 11.1-in. (282-mm) front, 11.4-in. (290-mm) rear, vacuum assisted

Swept area, sq in./sq cm	440	2839
Wheels		forged alloy, 16 x 7J
Tires		Pirelli P7, 225/50VR-16
Steering type		rack & pinion, power assisted
Overall ratio		17.8:1
Turns, lock-to-lock		3.1
Turning circle, ft/m	31.5	9.7

Front suspension: upper A-arms, lower trailing arms, coil springs, tube shocks, anti-roll bar

Rear suspension: upper transverse links, lower trailing arms, coil springs, tube shocks, anti-roll bar

INSTRUMENTATION

Instruments: 85-mph speedometer, 7600-rpm tach, 999,999 odo, 999.9 trip odo, oil press., coolant temp, voltmeter, fuel level, clock

Warning lights: central warning system, oil press., brake press., fluid, pad wear; parking brake; stop lamp, tail lamp failure; coolant, windshield-washer fluid, oxygen sensor, seatbeats, hazard, high beam, directionals

RELIABILITY

Owners of earlier-model Porsches reported 11 problem areas and 4 disabling reliability areas compared to overall Owner Survey averages of 12/7. So we expect the overall reliability of the Porsche 928 to be better than average.

ROAD TEST RESULTS

ACCELERATION

Time to distance, sec:

0-100 ft	3.5
0-500 ft	9.0
0-1320 ft (¼ mi)	16.2
Speed at end of ¼ mi, mph	89.5

Time to speed, sec:

0-30 mph	3.2
0-50 mph	6.3
0-60 mph	8.1
0-80 mph	13.0
0-100 mph	21.7

SPEEDS IN GEARS

3rd gear (5000 rpm)	est 140
2nd (6000)	107
1st (6000)	68

FUEL ECONOMY

Normal driving, mpg	est 18.0
Cruising range, mi (1-gal. res)	385

HANDLING

Lateral accel, 100-ft radius, g	0.811
Speed thru 700-ft slalom, mph	59.7

BRAKES

Minimum stopping distances, ft:

From 60 mph	138
From 80 mph	248
Control in panic stop	good
Pedal effort for 0.5g stop, lb	18

Fade: percent increase in pedal effort to maintain 0.5g deceleration in 6 stops from 60 mph ....nil

Parking: hold 30% grade?	yes
Overall brake rating	very good

INTERIOR NOISE

Idle in neutral, dBA	66
Maximum, 1st gear	83
Constant 30 mph	69
50 mph	72
70 mph	73
90 mph	78

SPEEDOMETER ERROR

30 mph indicated is actually	29.0
60 mph	59.5
80 mph	80.0

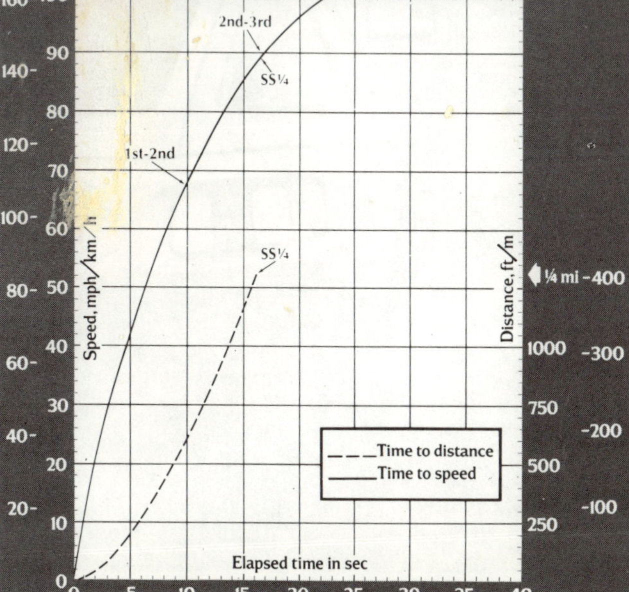

as power steering; more important to Porsche is the way it complements the 928's wonderfully balanced, capable handling with speed-sensitive assist to the rack-and-pinion steering gear.

No previous Porsche, in fact, has had such finely balanced handling as the 928—even though there are Porsches, like the late and lamented 930 Turbo, that can outdo it in terms of pure cornering power. It's the near 50/50 weight distribution and the engineers' advantage of starting from scratch with a new chassis that give the 928 its edge. Here's a sports car that gives almost race-car neutral response to the steering wheel—never mind that its driver is riding in the lap of luxury. It will oversteer if the driver so commands, but it takes either a very advanced, or very foolish, driver to push it that far. Most will simply enjoy the remarkably capable handling feel and the wide margin of safety between what they'll venture and what it can do.

We weren't quite so overwhelmed with the 928's brakes, even though they look just fine on paper. Under most conditions they do exactly what is required of them, but on a simulated panic stop from 80 mph the rear wheels tended to lock up and required considerable attention from the driver to produce minimum stopping distances. Given this, we averaged just 248 ft from 80 mph and, more relevant to driving in today's America, a very impressive 138 ft from 60 mph.

Speaking of 80 mph, it was a sort of cultural shock to settle into the 928, fasten the seatbelt, slip the selector lever into D and move out into traffic, only to find the speedometer needle getting vertical at city speeds. Sadly, even a machine with the capability of a 928 falls under the new U.S. "safety" requirement that speedometers read no more than 85 mph—a rule characterized by one voice within the National Highway Traffic Safety Administration itself as meaningless, but "cheap and easy." We expect that there will be a lively business from owners of 1980 and later Porsches (and makes of similar stature) in retrofitting "old-style" speedometers reflecting the machinery's true potential.

After all, how is the poor owner going to rationalize laying out $40,000-plus for a conveyance like this if he has to face up, every time he climbs behind the wheel, to the fact that a mere 85 mph is pretty much the outer limit in America today? In this sense, these are truly hard times for Porsche.

window. But we found the rear seats most useful with their backs folded flat, increasing the luggage capacity from 6.3 to 20.5 cu ft. And that would be a lot of space for a middle-size sedan, let alone a Porsche.

As time goes on, the tire engineers amaze us with their ability to combine lower and lower cross-sections and ever better handling with decent riding qualities. The 928's combination of 50-section Pirelli P7s (optional, as noted) and a painstakingly developed suspension system demonstrates further progress on this front. The tires, which appear to have virtually no height with which to absorb shocks and are inflated to 36 psi on the owner's manual's recommendation, are harsh over such sharp disturbances as lane-divider dots and tar strips and noisy on all but the smoothest asphalt surfaces (the latter attributable to their grippy tread design, not their proportions). But otherwise the ride is wonderfully supple and well controlled, at least for a decidedly high-caliber sports car.

The 928 is Porsche's first car with power steering, and this follows the practice established by other German carmakers of providing relatively little power assistance, a lot of road feel and a relatively direct ratio. It facilitates parking just enough to get by

Road Test Update

PORSCHE 928 AUTOMATIC

What price perfection?

WHEN THE PORSCHE 928 was introduced early in 1977, we said, "Tomorrow is here." You see, we had just had our first taste of this revolutionary GT and were impressed. We were also confident that the enthusiast public would be too, once they had experienced the 928. Well, it has been more than three years since that glorious moment, yet the enthusiasts seem largely unimpressed by a car that is light years ahead of almost every other automotive design, technologically, ergonomically and stylistically. Only the cognoscenti (who are few and far between) and our staff (who know a good thing when they see one) continue to be impressed by this GT that one of our staffers calls, "an engineering tour de force." His comment is mirrored by the glowing praise of another staff member who says, "The 928 is well-nigh absolute perfection."

So what's the problem? Perhaps it's that the 928 is so perfect and so easy to drive that it doesn't exude the sort of mechanical machismo that makes such cars appealing. Or it may be that there is some buyer resistance to a car that costs $10,000 more than the 911SC, which many purists consider to be the only real Porsche. Some dealers refute this last belief, saying that if you can sell a Porsche buyer a 911SC, you can just as easily sell him a 928.

But if that's true, then why do the air-cooled cars outsell the water-cooled ones, two-to-one? Could it be that the 911SC best fits the Porsche image?

Whatever the reason, anyone who appreciates sophisticated engineering and design, not to mention superb handling, performance, braking and comfort does himself an injustice by not driving a 928. "The only thing I don't like is the price," said one tester, who claimed the 928 is the one car he'd buy if his financial situation were rosier. As it is, the car's $38,000 price tag scares off our staff and, we suspect, most middle-income buyers.

With the demise of the 930 Turbo in this country, the 928 has become the flagship of the Porsche fleet. And an impressive craft it is. Leather is standard for 1980. "It delights the touch and the nose," commented one driver. Almost as delightful are such standard items as air conditioning, cruise control, electric window lifts, central door locking and an AM/FM/stereo-cassette, all carryovers from last year. Alas, the headlight washers, high-pressure windshield washers and rear-window wiper, formerly standard equipment, are now packaged as a $275 option. Other options include an electric sunroof, climate control, electric seat adjustment and 16-in. wheels with Pirelli P7 tires. The good news for those purists who demand a bit more excitement from their

a substantial increase in torque: now 265 lb-ft at 3600 rpm, an 11 lb-ft gain over the 1978 model. How do the new emission controls affect driveability? Frankly, we hardly noted any change because previous models were exemplary in that respect. Fuel consumption is something that can be measured and as expected, the new controls do help, boosting mpg from 15.0 to 16.5.

The Mercedes-built 3-speed automatic is a no-extra-cost alternative to the standard 5-speed. This torque-converter-equipped gearbox is well matched to the Porsche V-8's output, and its normal soft, smooth shifts get progressively crisper and firmer the longer you keep your throttle foot buried in the floorboard. A proper compromise we think. Optimum shift points occur at about 5500 rpm, even though the engine is capable of pulling smoothly to its 6000-rpm redline; and with a minimum of audible protestation, we might add.

Driving a 928 is always a pleasurable experience. The ride is firm yet compliant and the P7s provide balanced handling and excellent high-speed stability. The variable-boost steering has excellent feel. It's not twitchy nor overly heavy. The all-disc braking system is more than a match for a car of the 928's speed potential and during spirited driving the brakes really come into their own. They are nearly impossible to lock from high speeds, which is what's really important.

As much a part of driving the 928 is enjoying its plush cabin and letting the little ergonomic details tickle your fancy—things like the adjustable steering wheel/instrument cluster, the form-fitting seats, the easy-to-reach controls, the dead pedal, the expandable map pockets in the armrests, the coin holder for expressway tolls and the sunshades for rear-seat passengers.

At $37,930, this superb Gran Turismo has the dubious honor of enjoying exoticar status. But the 928 is more than just an expensive exotic. A Ferrari might have more snob appeal, but for all-around use, we don't think you can top the Porsche 928. ⊘

machinery is that Porsche will offer an S version of the 928. Equipped with front and rear spoilers, heavy-duty shocks, a limited-slip differential, forged alloy wheels and sport seats, it will be the sort of car they expect from Zuffenhausen. The bad news is it won't be available until the 1981 model year.

Not that the present 928 needs much improvement. For instance, its performance is already impressive, even with an automatic transmission, which is what we tested this time around. From a standing start, the 1980 model reached 60 mph in 8.1 seconds and covered the quarter mile in 16.1 sec, 0.2 and 0.5 sec quicker than its similarly equipped 1978 predecessor. The secret lies under the bonnet where the 4474-cc sohc V-8 now sports a 3-way catalyst with oxygen sensor emissions package that offers an infinitesimal horsepower gain (220 versus 219 bhp), but

PORSCHE 911SC

The more things change, the more they remain the same

PHOTOS BY JOE RUSZ

EACH YEAR, DURING one of our many conversations with Porsche factory representatives, we invariably ask, "How long will you continue building the 911?" And each year the answer is the same: "As long as the demand for the car continues and it is able to meet safety and emissions standards." We have accepted this reply with a bit of skepticism and the recent demise of the U.S. version of the 930 Turbo feeds our suspicions. It disappeared from our market place without even an *auf Wiedersehen* because Porsche no longer found it practical to build an American version of this sensational GT. But the 911 appears to be a different matter entirely and even though we were convinced that the introduction of the water-cooled, front-engine 924 and 928 signaled the end of the air-cooled, rear-engine era, Porsche has proved us wrong. We largely ignored the factory's words and failed to reckon with the continuing popularity of the 911. In fact, such is the demand for this car, that two out of every three models built in the Zuffenhausen assembly plant are 911s.

If you have not driven the latest 911, officially called 911SC, since 1978, perhaps you should. Then you'll understand what all the adulation is about. You might even be prompted to make the sort of remark one of our freshman staff members made after driving our test car: "Now I know why the 911 has been so popular for so long. It is well built, esthetically pleasing, comfortable and handles and performs very well."

Porsche is not one to make wholesale changes to a successful design although there have been numerous improvements in the 17 years since the 911's introduction. The wheelbase has been stretched, the fenders flared, the ventilation system and interior modernized. Yet the car still looks like its predecessors. But only the engine displacement has undergone a substantial increase: The flat-6 has grown from its original 1991 cc in 1963 to 2994 now. Porsche says this is the maximum for the production engine, though Director of Public Relations and Sport Manfred Jantke drives a 911SC with a specially built 3.3-liter engine. You don't

need one, claims the factory, and they are probably right. The 1980 version, which develops the same 172 bhp and 189 lb-ft of torque as the 1978 and 1979 models, has a top speed of 139 mph, according to Porsche.

For 1980, the significant news is that the latest 911SC is now a 50-state car. It's fitted with a 3-way catalytic converter and when this is used in conjunction with the existing Bosch K-Jetronic fuel injection, the exhaust gases are clean enough to pass even California's stringent pollution laws—without the use of EGR. This new combination also allows the use of a higher compression ratio—9.3 versus 8.5:1—and the result is crisper throttle response, better driveability and improved fuel economy.

While we're on the subject, here's what else is new for 1980: options or the lack of same. For example, the most popular add-ons are now standard: electric window lifts, air conditioning, a ⟫⟫→

<table>
<tr><td colspan="4" align="center">**AT A GLANCE**</td></tr>
<tr><td></td><td>Porsche
911SC</td><td>Porsche
924
Turbo</td><td>Ferrari
308 GT4</td></tr>
<tr><td>List price</td><td>$27,700</td><td>$20,875</td><td>$38,460</td></tr>
<tr><td>Curb weight, lb</td><td>2805</td><td>2835</td><td>3405</td></tr>
<tr><td>Engine</td><td>flat 6</td><td>inline 4</td><td>V-8</td></tr>
<tr><td>Transmission</td><td>5-sp M</td><td>5-sp M</td><td>5-sp M</td></tr>
<tr><td>0-60 mph, sec</td><td>6.7</td><td>7.7</td><td>7.8</td></tr>
<tr><td>Standing ¼ mi, sec</td><td>15.3</td><td>16.3</td><td>16.0</td></tr>
<tr><td>Speed at end of ¼ mi, mph</td><td>91.0</td><td>88.0</td><td>89.0</td></tr>
<tr><td>Stopping distance from 60 mph, ft</td><td>140</td><td>na</td><td>167</td></tr>
<tr><td>Interior noise at 50 mph, dBA</td><td>72</td><td>70</td><td>77</td></tr>
<tr><td>Lateral acceleration, g</td><td>0.798</td><td>na</td><td>0.779</td></tr>
<tr><td>Slalom speed, mph</td><td>59.7</td><td>na</td><td>57.9</td></tr>
<tr><td>Fuel economy, mpg</td><td>18.5</td><td>est 21.0</td><td>14.0</td></tr>
</table>

leather-covered steering wheel, center console, black window trim and engine compartment light. Equipped in this fashion, the basic 911SC costs $25,900, quite a hefty piece of change. The Targa, a separate model, costs $1350 more than the coupe, but it offers the sort of open-air motoring the romantics among us were weaned on. One personal observation by our resident Porsche expert, Joe Rusz, is that Targas suffer from excessive wind noise, even with the top in place, at speeds in excess of 100 mph.

What we said about the 1978 911SC (R&T, April 1978), generally goes for the 1980 version. Handling is exceptional: 0.798g on the skidpad, 59.7 mph in the slalom (thanks to those optional 6- and 7-in. wide wheels and Pirelli P7 tires). Yes, as we said then, they are "thumpy over tar strips and lane divider dots and noisy over most surfaces." But it's a small price to pay (figuratively) for the handling security this package offers. Because in case you've forgotten, the 911 has extreme rear weight bias and thus is prone to oversteer. Wide wheels and ultra-sticky

tires are one way of taming this condition. As it is, some drivers complained of twitchiness, especially in the rain. This may have been partly because of mild hydroplaning which one expects from unusually wide tires. The car also displayed crosswind sensitivity, but that is the nature of a rear-engine beast. The Editor summed up the situation nicely when he said: "In very high-speed driving the 911SC still has characteristic, uncomfortable front-end lightness. And it still has characteristic off-throttle oversteer and on-throttle understeer that has been tamed a fair bit by those wide wheels and P7s." Which, we might add, look just right on the 911SC.

The 911SC's interior is similar to that of earlier 911s. Or, as one staffer said, "If you've been in one 911, you've been in them all." He's right, because in a 911, the primary instruments and controls are always in the same place. For instance, the tachometer is prominent, the speedometer almost secondary. Most enthusiasts like to drive by the tach anyway. This year the layout seems to make even more sense because Porsche, in complying with U.S. standards, has installed an 85-mph speedometer—in a 140-mph road car? As for the rest of the controls and gauges, well, you know the drill. The shifter falls readily to hand; the steering wheel is perfectly positioned for assuming the standard 9 o'clock-3 o'clock grip. Everything vital to driving is close and convenient, except for one unpardonable sin: the brake/accelerator pedal positioning. In recent years, Porsche has changed the height relationship between the two so that it's impossible to heel-and-toe with any sort of alacrity. The solution to this annoying problem is simple if you don't mind spending a few minutes playing with a few wrenches. The SC's gas pedal stop is adjustable, as is the cable, allowing you to raise or lower the pedal height while maintaining pedal travel. Also, the upper stop on the brake pedal is adjustable and you can move this stop closer to the floor while shortening the threaded master cylinder pushrod to maintain the same travel relationship. In Germany the throttle pedal is adjusted so that the driver is comfortable with his foot to the floor. In deference to typical U.S. driving conditions, the gas pedal is adjusted closer to the floor to match what Porsche considers the ideal foot/leg angle when the throttle is just cracked open for 55-mph cruising.

Driveability of Porsche's venerable flat-6 engine with its 3-way catalyst emission control system is excellent; the power, more than sufficient. There are no peaks, valleys or flat spots in acceleration, and the engine is flexible enough so you can lug it to 1000 rpm in 5th gear, without any mechanical protestation. Not that any true Porsche owner will ever let his engine see less than 2000 rpm in any gear! Once it's revving—and it does so freely, pulling especially strong from about 3500 rpm—we saw 138 mph at 6000 rpm in 5th gear, and the engine was still pulling. Incidentally, the top speed shown for the 1978 SC test in the 1978 issue is incorrect; it should have been 136, not 126 mph.

Early in our test period, some drivers complained about the 911SC's cold-start problems. The car had a tendency to stall repeatedly, even though it had been garaged overnight. But about two weeks later the problem disappeared and we suspect that the car's newness (it had 772 miles on the odometer upon delivery) was to blame.

Performance has always been the 911SC's forte and factoring in the car's lack of break-in miles, a 60-lb heavier test weight and horsepower and torque figures identical to our 1978 SC, acceleration times are just about the same. We measured 0–60 mph in 6.7 seconds and clocked the quarter mile in 15.3 sec at 91.0 mph.

Fuel economy is also about the same: 18.5 mpg versus 18.0 in 1978. We expected the 3-way-catalyst-equipped engine to do better, but this may be partly a function of every driver making frequent use of the 911SC's impressive performance.

Two years ago we said, "There are many cars more exotic than a 911, but few are as practical and as thoroughly usable." Our opinion hasn't changed. In fact, it has been reinforced by the latest accolades of our staff. "There's no other car that offers the unique blend of characteristics found in the 911," commented one editor. "I'd still vote it one of the best designs of the Eighties." Another said, "The 911 is still a highly desirable sports/GT with comfort, handling and, most importantly, performance."

Unfortunately, most of us also agree that at $30,000, the average price of 911SCs being sold today, this Porsche is beyond the reach of most middle-income car buyers. Our Editor said, "If you had told me five years ago that in 1980 the 911 would be an exoticar, at least in price, I'd have laughed." He's not laughing, nor are the rest of us. But for those who can afford to chuckle at the cost, the 911SC is no laughing matter. In fact, you'll find that as cars become increasingly anemic, the 911 shines brighter.

PRICE

List price, all POE	$27,700
Price as tested	$30,935

Price as tested includes standard equipment (air cond, elect. window lifts), power antenna ($195), elect. sunroof ($895), Option F25: forged alloy wheels & P7s, elect. right-hand mirror, 2 additional rear speakers, sport shocks, fog lights ($2145)

GENERAL

Curb weight, lb/kg	2805	1273
Test weight	2950	1339
Weight dist (with driver), f/r, %		41/59
Wheelbase, in./mm	89.4	2271
Track, front/rear	53.6/52.8	1361/1341
Length	168.9	4290
Width	65.0	1651
Height	52.0	1321
Trunk space, cu ft/liters	4.5	127
Fuel capacity, U.S. gal./liters	21.1	80

ENGINE

Type	sohc flat 6
Bore x stroke, in./mm	3.74 x 2.77 ....95.0 x 70.4
Displacement, cu in./cc	183 2994
Compression ratio	9.3:1
Bhp @ rpm, SAE net/kW	172/128 @ 5500
Torque @ rpm, lb-ft/Nm	189/256 @ 4200
Fuel injection	Bosch K-Jetronic
Fuel requirement	unleaded, 91-oct

DRIVETRAIN

Transmission	5-sp manual
Gear ratios: 5th (0.82)	3.18:1
4th (1.00)	3.88:1
3rd (1.26)	4.89:1
2nd (1.83)	7.10:1
1st (3.18)	12.34:1
Final drive ratio	3.88:1

CHASSIS & BODY

Layout	rear engine/rear drive
Body/frame	unit steel
Brake system	vented discs; 11.1-in. (282-mm) front, 11.4-in. (290-mm) rear; vacuum assisted
Wheels	forged alloy; 16 x 6J front, 16 x 7J rear
Tires	Pirelli P7; 205/55VR-16 front, 225/50VR-16 rear
Steering type	rack & pinion
Turns, lock-to-lock	3.1

Suspension, front/rear: MacPherson struts, lower arms, torsion bars, tube shocks, anti-roll bar/semi-trailing arms, torsion bars, tube shocks, anti-roll bar

CALCULATED DATA

Lb/bhp (test weight)	17.2
Mph/1000 rpm (5th gear)	23.2
Engine revs/mi (60 mph)	2590
R&T steering index	1.04
Brake swept area, sq in./ton	338

ROAD TEST RESULTS

ACCELERATION

Time to distance, sec:

0–100 ft	3.1
0–500 ft	8.2
0–1320 ft (¼ mi)	15.3
Speed at end of ¼ mi, mph	91.0

Time to speed, sec:

0–30 mph	2.2
0–50 mph	4.9
0–60 mph	6.7
0–70 mph	9.0
0–80 mph	11.5
0–100 mph	19.7

SPEEDS IN GEARS

5th gear (6000 rpm)	138
4th (6500)	123
3rd (6500)	90
2nd (6500)	64
1st (6500)	36

FUEL ECONOMY

Normal driving, mpg	18.5

BRAKES

Minimum stopping distances, ft:

From 60 mph	140
From 80 mph	248
Control in panic stop	fair
Pedal effort for 0.5g stop, lb	32

Fade: percent increase in pedal effort to maintain 0.5g deceleration in 6 stops from 60 mph ... nil

Overall brake rating ... very good

HANDLING

Lateral accel, 100-ft radius, g	0.798
Speed thru 700-ft slalom, mph	59.7

INTERIOR NOISE

Constant 30 mph, dBA	70
50 mph	72
70 mph	76

SPEEDOMETER ERROR

30 mph indicated is actually	27.0
60 mph	59.0

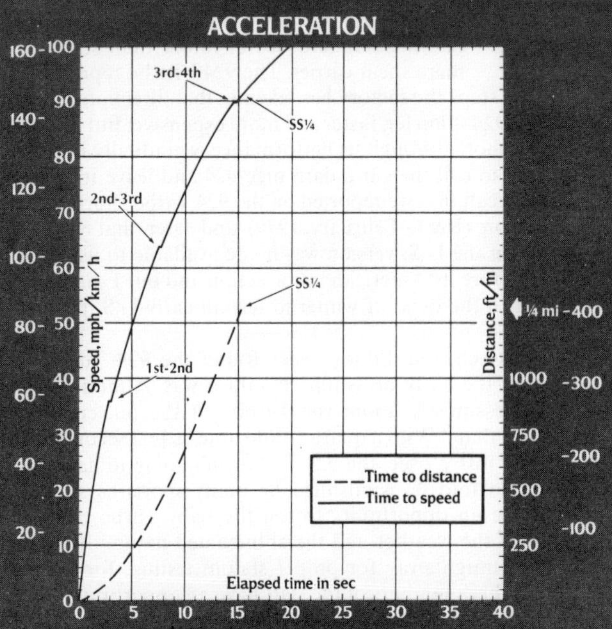

1980 PORSCHE 924S TURBO

Pointing the way to Porsche's future

PHOTOS BY JOE RUSZ

PORSCHE HAS GREAT plans for the 924. The car that began life as a Volkswagen, then became the basic model for the Porsche lineup, is slated to move up-market until one day it will replace the 911 as the Porsche standard bearer rather than serving as a mere spear carrier. The 924S Turbo represents one of the first steps the factory has taken in that direction. Call it an improved 924—fancier, faster and more expensive. But a 911? Not by a long shot, although its performance is gradually improving. We prefer to call the car a darn nice 924 and leave it at that.

You'll recall that we reported on the 924 Turbo shortly after its introduction (R&T, February 1979) and later that year (June 1979) when the U.S. version was made available to us. This was months before its American introduction and our Editor went to Weissach in the dead of winter to test an early U.S. production model.

At Weissach, our Editor/tester found the 924 Turbo to be everything the early press reports said it was. Although he was "mildly pessimistic about the Turbo" at the outset, he soon discovered that, "With a quarter-mile time of 16.3 seconds and 0–60 mph in just 7.7 sec, the 924 Turbo isn't going to have rubber dust kicked into its windshield by many sports cars." But he didn't have an opportunity to test the baby Turbo thoroughly because of the weather and the abbreviated nature of Weissach (no long straightaway for proper slalom testing, for instance). Nor did other members of the staff have an opportunity to share this excitement.

We decided to remedy this situation in a big way, by testing a 1980 model that's one step closer to Porsche's goal of making the 924 line a worthy successor to the 911. No ordinary 924 Turbo this, but one with the S package, a $2045 collection of handling/braking/appearance features that some might argue belong on a car whose base price runs $20,875 in the first place. This Sport Group, for instance, includes ventilated disc brakes on all four

<table>
<tr><td colspan="4" align="center">AT A GLANCE</td></tr>
<tr><td></td><td>Porsche
924S
Turbo</td><td>Chevrolet
Corvette</td><td>Porsche
911SC</td></tr>
<tr><td>List price</td><td>$22,920*</td><td>$13,140</td><td>$25,900</td></tr>
<tr><td>Curb weight, lb</td><td>2850</td><td>3345</td><td>2805</td></tr>
<tr><td>Engine</td><td>inline 4</td><td>V-8</td><td>flat 6</td></tr>
<tr><td>Transmission</td><td>5-sp M</td><td>4-sp M</td><td>5-sp M</td></tr>
<tr><td>0–60 mph, sec</td><td>9.3</td><td>7.7</td><td>6.7</td></tr>
<tr><td>Standing ¼ mi, sec</td><td>17.0</td><td>16.0</td><td>15.3</td></tr>
<tr><td>Speed at end of ¼ mi, mph</td><td>81.0</td><td>86.5</td><td>91.0</td></tr>
<tr><td>Stopping distance from 60 mph, ft</td><td>155</td><td>130</td><td>140</td></tr>
<tr><td>Interior noise at 50 mph, dBA</td><td>69</td><td>73</td><td>72</td></tr>
<tr><td>Lateral acceleration, g</td><td>0.827</td><td>0.790</td><td>0.798</td></tr>
<tr><td>Slalom speed, mph</td><td>61.5</td><td>61.2</td><td>59.7</td></tr>
<tr><td>Fuel economy, mpg</td><td>20.0</td><td>14.5</td><td>18.5</td></tr>
<tr><td colspan="4">*Price shown includes Sport Group ($2045).</td></tr>
</table>

corners (unlike the standard U.S. Turbo's rear drums) and 5-bolt forged alloy wheels fitted with Pirelli P7 tires instead of the 4-bolt cast wheels cum Pirelli CN36s found on ordinary 924 Turbos. Stiffer shocks, a fatter front anti-roll bar (23-mm diameter versus the standard 21 mm) and a 14-mm bar added to the rear round out the suspension side of the package.

The engine for the 924 Turbo, S or otherwise, is the same K-Jetronic-injected, KKK-turbocharged powerplant found in the 1979 model. It's a 50-state engine whose 3-way catalyst and unleaded 91-octane diet still allow it to produce 143 bhp at 5500 rpm and 147 lb-ft of torque at 3000. This is 28 bhp more than that of the normally aspirated 924, so you'd expect the Turbo to be more than a bit quicker. And it is, although off-the-line performance of the car we tested actually fell between figures for the normally aspirated car and for the early production Turbo we tested at Weissach. For instance, our stateside 924S Turbo went from 0 to 60 mph in 9.3 sec and zipped through the quarter mile in 17.0 sec at 81.0 mph. No slouch indeed, but pretty far off the breathtaking kick of the Weissach car. We've done some checking, talked to people at Porsche-Audi and elsewhere and we've come to recognize several factors that combine to explain the differences. For one, our S-optioned car carried Pirelli P7s, stickier and wider than the CN36s fitted to the Weissach car, and the P7s' added grip meant they didn't light quite so readily off the line. Second, Orange County International Raceway's surface is different from Weissach's, and there were wildly different ambient temperatures during Weissach's winter and OCIR's summer. Third and perhaps the most important, the Weissach car was an early production version calibrated to 1979 U.S. emission standards, whereas our latest 924 Turbo was a 1980 model brought through the ordinary pipeline. Last, we can note that the factory's own literature gives 9.3 sec as the 0–60 mph time for the U.S. 924 Turbo, and based on our experience with several of the cars, we're convinced this performance is typical.

All this has a certain irrelevance, though, when you experience the tractability and free-revving nature of the 924 Turbo's powerplant. The engine will pull away from as low as 1000 rpm or it can be twisted as high as 6200 until its rev limiter comes into action—curiously enough, 400 rpm below the tach's redline. Turbo boost can be felt as low as 2000 rpm, which means that one can handle traffic without a lot of fancy footwork.

A good thing too, because our car's shift linkage was not the best. Several staff members found it balky and said the gates were indistinct in its 2nd/4th definition. It's a 5-speed gearbox, with 1st left and back next to your right leg and reverse directly above it. Several staff members complained about what Porsche calls this "racer pattern," and said they would prefer the more common setup with 1st at the top of the H and 2nd below it. We asked Porsche about it and the explanation was that the original 924 gearbox was designed this way (the factory considers 1st to be

a largely unused gear, it seems). Since the Turbo box is based on this original one, it has the same shift pattern. But then why do normally aspirated 1980 924s use a conventional 5-speed pattern? Because that new lighter-duty design is essentially a 4-speed with 5th tacked on.

Other elements of the 924 Turbo are generally unchanged from the 1979 model. This is good and bad. The good news is that many amenities like electric window lifts are still standard. The bad news is that the steering wheel still brushes against one's thigh when it's turned more than 90 degrees in either direction. It's a problem we've encountered with all 924s, and while the removal of four spacers in the seat mounting can help matters (see our 924 road test, April 1980), it leaves the seat lower than some might like. There's also an optional 4-spoke wheel with more axial offset and smaller diameter, yet we're puzzled

CONTINUED ON PAGE 55

Road & Track Owner Survey

PORSCHE 911

Pricing itself out of the market?

ILLUSTRATION BY BILL DOBSON

PORTS AND GT cars were formerly considered temperamental but worth it for the driving fun they provided. In recent years, however, many of them have become very sophisticated pieces of engineering with high price tags. Such a car is the Porsche 911. Curiously, it's also a car that has been generally surrounded by a staunch group of disciples who must fend off the criticisms of the detractors; in other words, some folks love Porsche 911s and others can't stand them. We expect that most of those who attack the 911 have never owned one, and perhaps have never spent much time driving one. But the purpose of this survey is to reveal how those who have owned and lived with 911s feel about them, as well as point out facts regarding dealers' service, maintenance and problem areas. This report covers 149 cars from 1974 through 1978 models. Considering the tremendous escalation in price that has taken place in the past several years, many readers who may not be able to afford a new 911 might want to use this survey as a tool for buying a used 911.

Of the 149 Porsches that make up the survey, 21 (14%) are 1974 models, 25 (17%) date from 1975, 28 (19%) from 1976, 40 (27%) 1977, and 35 (23%) are of 1978 vintage. All were bought new (we no longer consider cars for surveys that are purchased used because there's no way of knowing what the original owner may have done to the car) and just more than half (52%) had between 10,000 and 20,000 miles on them. Fifteen percent fell into the

20,000–30,000-mile range and 16% between 30,000 and 40,000, with the remainder having been driven more than 40,000 miles. We would expect Porsche owners to be relatively affluent people and that is the case with this group if a judgment can be made on the basis of multiple car ownership: only 19% are single-car owners (compared to 42% of the Triumph TR7 owners surveyed in June 1980), while 46% own some other car in addition to the Porsche, 24% claim three cars, 7% have four, 3% have five and 1% own six or more.

A mark of Porsche dealers' marketing philosophy is evidenced by the addition of optional equipment and Porsche buyers aren't bashful in this regard: 62% added air conditioning, 72% wanted AM/FM/tape player sound systems, 50% opted for leather upholstery, 65% paid for special wheels, 40% bought sunroofs, and 42% sought various other options ranging from fancy steering wheels to headlight washer systems. Two of the 149 owners selected automatic transmissions as an option and, from 1976 on, the standard manual gearbox has been a 5-speed. For 1974 and 1975, 64% of the owners opted for the 5-speed while the others contented themselves with the 4-speed. Five of the 1976 models are Turbo versions and one 1978 car is too, but we found nothing in the owners' responses to differentiate these cars from the normally aspirated 911s in any way. So, the remaining 143 cars are all equipped with the 2.7-liter engine and K-Jetronic fuel

injection introduced with the 1974 model year.

Who Bought Them & Why

W E EXPECTED most Porsche 911 owners would base their purchase decision on such factors as handling, fun to drive and performance—and we were correct. Handling was cited by 95% as a feature that influenced their choice of the 911, 94% mentioned fun to drive, and 94% liked the performance. Most people think of Porsches as having sophisticated engineering and 95% of our respondents said that too was a factor, while 90% indicated Porsche's reputation for quality played a role in their purchase of the car. Eighty-four percent also mentioned workmanship, 68% cited the reputation for reliability and 59% liked the styling.

While most of the cars we've surveyed are driven daily by more than 90% of the owners, only 81% of the 911 owners use their cars for daily transportation which probably reflects the high level of multi-car ownership. Befitting the 911's GT nature, 58% of the owners use them for vacations and long trips, and quite a number (22%) enter their Porsches in rallies and/or slaloms. Also, 11% use them for other activities, from racing to weekend jaunts.

Considering the sporting characteristics of the 911, it's not surprising that only 23% describe their driving as moderate, 64% say they drive hard (well above average) and 13% checked very hard on the questionnaire. The fact that 911 owners demand a lot from their cars on the road doesn't in any way mean they abuse them, however, as a whopping 46% say they do more maintenance than called for by the manufacturer's schedule. Forty-four percent said they follow the manufacturer's maintenance schedule, 9% claimed to follow it mostly, and only 1% admitted they don't keep up their cars as they should. In addition to pride of ownership and a desire to have the 911 running properly, the fact that Porsche owners have a substantial financial investment in their cars probably also helps account for the unusually high level of maintenance.

In keeping their cars in proper working order, Porsche owners are by and large pleased with the service they receive from dealers: 34% rate dealers' service excellent and 27% give it a good rating for a combined total of 61%, which is somewhat better than the average of all the other cars we've surveyed. Nineteen percent rate Porsche dealer service fair and 20% give it a poor rating. Among the latter two groups, many complained of the high prices for parts and labor, others were upset by what they considered incompetence, especially when they had to take their cars back two and three times to get them fixed properly, and some felt they were dealing with prima donnas who acted like they were doing the customer a favor by working on the car. On the other side, however, some owners were quite pleased with dealer service, such as the Alabama 911 owner who noted his dealer is "Eager to help with any problems that arise," or the Maryland owner who wrote that the mechanics have good knowledge of the car, are courteous and willing to explain technical features to him.

Regarding our relatively new question about the car being out of commission while awaiting parts, 28% of the 911 owners said yes. The average for the six cars on which we've tabulated this question is 24%, so the 911 is slightly worse than average in this regard. Curiously, there's no apparent year-by-year pattern, as 1974, 1975 and 1977 cars had a fairly high rate of waiting for parts, while 1976 and 1978 models didn't. The most frequently mentioned parts, incidentally, were associated with the K-Jetronic fuel injection.

Best & Worst Features

M OST PORSCHE 911 owners say their cars live up to their expectations. In fact, the features that influenced their choosing the 911 were the same ones checked in the Best Features category after having driven the car for 10,000 miles or more: Handling (96%), fun to drive (91%), performance (87%), engineering (85%), workmanship (75%), overall quality (72%), reliability (55%) and styling (51%). Almost half (44%) of the owners feel comfort is a best feature, and a third (33%) appreciate the ride characteristics. Twenty-five percent note fuel economy as a best feature and nearly one of every five owners (19%) checked economy of operation.

Looking at what Porsche owners don't like, 23% mention the lack of space, 17% object to the noise level, 13% complain about the inadequate ventilation, 12% find the cost of parts and service too high, and 7% find poor fuel economy and lack of reliability worst features. Six percent feel the 911's price tag is exorbitant, but on the other hand, 16% of the owners couldn't come up with a worst feature at all—many of them noted that it's simply "the perfect car."

Problem Areas

W E SURVEYED Porsche 911/912 owners in February 1969 and that group reported 8 problem areas shared by 5% or more of the owners. In the present survey of 911s, 12 problems are mentioned, with 5 shared by 10% or more of the respondents, and 7 afflicting 5–10% of the cars. The largest number, 21%, goes to miscellaneous engine problems, primarily the oil pump and sealing ills and chain tensioners; 19% goes to the miscellaneous electrical category and the culprits here were mostly the sunroof and the air conditioner. Eighteen percent of the owners reported fuel system-related ailments and 12% had troubles with various instruments. Porsche has made some progress in this area, because 22% of the 1969 owners surveyed had instrument problems. It's also noteworthy that an astounding 40% of the 1969 owners reported sparkplug troubles, while we found only a single case of serious sparkplug difficulties in the latest group. Ten percent of the 911s in this survey suffered alternator/voltage regulator ills, which are so closely related we combined them into a single group.

In the 5–10% area, 9% had problems with body parts, front brakes, clutches and exhaust systems; 7% cited window lifts and shock absorbers; and 6% complained of premature tire wear.

We found no surprises in the problems experienced by this group of Porsche owners. Sunroofs and air conditioners are not common options in the European market, so it's not too startling ⟫⟫⟫

that Porsche has a tough time making them foolproof. (We found similar problems in our August 1977 survey of Mercedes-Benz 450 models, but as Porsche sells roughly half of all 911s in the U.S. market, they could be expected to do better.) The same can perhaps be said of electric window lifts. Exhaust system ills have been a Porsche tradition for years, as the heater boxes just don't hold up, especially as emission controls produce rising temperatures in the exhaust system. Porsches have also shown a marked tendency to body rust since the 911 model was introduced in 1965 but significant progress has been made in this area in recent years and the rust problems mentioned here are primarily found on the older cars in this survey. On the plus side, 13% of the 911 owners reported no problems with their cars.

Buy Another?

FOR ALL the cars we've surveyed since 1968, the average number of owners who are sufficiently pleased with their cars to buy another of the same make is 75%—for the Porsche 911 it's 88%, and that's quite good. Not many marques or other models can boast a higher percentage than that—the Porsche 914 (November 1974) registered 89%, the Porsche 911/912 survey of 1969 notched 95%, and the Alfa Romeo Giulias (January 1970) and Honda Civic (July 1975) both garnered 94%. The current group of 911 owners are clearly people who are enthusiastic and involved with their automobiles: "The car delivers on command—incredible brakes, excellent cornering and very fast," wrote a professional figure skater from Allentown, Pennsylvania;

"It's a super car!" was the conclusion of a doctor from Anacortes, Washington. And a Silver Spring, Maryland owner suggested that the price be lowered "so I can buy one for my son who drives me crazy wanting to drive my car. Of course I cannot permit him this luxury."

On the other side of the "buy another" coin, 5% say they would not. Most of these owners felt they had had too many problems with a car costing so much and they simply couldn't justify the purchase of another: "You don't get as much for your money as might be had in other makes," wrote a Fargo, North Dakota owner.

Another 7% of the 911 owners are undecided about buying another Porsche, but quite a few of them have simply been priced out of the market and while they enjoy their present one, don't feel they can afford a new one.

The Porsche 911 has been a benchmark sports/GT car since its introduction some 15 years ago—the sort of car by which others are measured. Many of the survey respondents offered pleas that we try to influence Porsche to continue producing the 911, which has been rumored to be on the way out for the past several years. Sales have remained healthy, however, and as long as Porsche can certify the 911 for U.S. safety and emissions, the company will continue to offer it for sale.

In our survey of these 1974–1978 models, we've discovered that it's not a perfect car—12 problem areas is one more than average—but one that offers a high degree of driving pleasure. And in the world of sports and GT cars, that's worth a lot.

SUMMARY: PORSCHE 911

	Porsche 911	Average for all Surveys since 1975
How Driven		
Moderately	23%	36%
Hard	64%	51%
Very Hard	13%	13%
How Owners Rate Dealer Service		
Excellent	34%	27%
Good	27%	31%
Fair	19%	21%
Poor	20%	21%
Maintained by the Book?		
Yes	44%	57%*
No	1%	8%
Mostly	9%	19%
More than recommended	46%	34%

*Surveys until May 1977 included those who did more than recommended maintenance in this category too.

Buy Another of the Same Make?	Porsche 911	Average for all Surveys since 1975
Yes	88%	75%
No	5%	17%
Undecided	7%	8%

Problems:	Porsche 911	Average for all Surveys since 1975
Common to 10% or more	5	7
Instruments		
Miscellaneous electrical (sunroof & air conditioning)		
Fuel pump/fuel injection*		
Miscellaneous engine* (chain tensioners, oil seals)		
Alternator/voltage regulator*		
Common to 5–10%	7	5
Body parts		
Front brakes		
Clutch*		
Window lifts		
Exhaust system		
Shock absorbers		
Tires		
Affecting reliability	4	4

*Represents a reliability area that could make the car unsafe or impossible to drive.

Averages from all Previous Surveys		
Problem areas		11
Reliability areas		4

Total Mileage on Car Now

10,000–20,000	52%
20,000–30,000	15%
30,000–40,000	16%
40,000–50,000	11%
50,000 +	6%
(Median	19,600)

Five Best Features
Handling
Fun to drive
Performance
Engineering
Workmanship

Five Worst Features
Lack of space
Noise
Cost of parts & service
Ventilation
Fuel economy, Reliability (tie)

CONTINUED FROM PAGE 51

why Porsche doesn't fix it properly by moving the steering column upward a tad. It would solve the thigh problem, and instrument sight lines would be better too.

One positive note on the current steering wheel—its 3-spoke design offers an excellent grip, which is handy when driving the 924 Turbo in its proper manner—briskly. Most of the car's negative traits are overlooked when the Turbo is flung down some twisty highway at a rapid clip. That's when the suspension, those sticky P7s, incredible brakes and flexible engine work together to produce the excellent balance one hopes for, but doesn't always encounter, in many expensive GTs. True, there's a tendency toward mild understeer, but this can be modulated by backing off the throttle slightly. The P7s are predictable and build up cornering power at a linear rate. The brakes are extremely fade-resistant, and slow the car in a straight line, stop after stop. In our instrumented testing, for instance, panic-stop distances were 155 ft from 60 mph and 258 ft from 80, control was very good, and brake fade was nil in six 0.5g stops from 60.

With the sport shocks, the ride is firm and, to one test driver, very reminiscent of the 911. This may be intentional, since, as we said, the 924 is being groomed as the 911's successor. One drawback is the steering, which is heavy enough at low speeds to mask the P7s' excellent characteristics.

Of course, the 924 Turbo's on-road handling is a relatively subjective sort of thing and you'll recall that, earlier, we hoped to learn how the car performed in a controlled environment— through the slalom and on the skidpad. If you guessed, "sensationally," you're correct. With its P7s and those 6-in.-wide Bonneville streamliner wheels, the 924S Turbo weaved around the cones at a speed of 61.5 mph, considerably quicker than the standard 924 which did 58.9 mph. On the skidpad, the Turbo delivered more of the same excellent handling, achieving 0.827g lateral acceleration, which made it 4th (or 3rd, if you consider 1st a tie) in our all-time standings. The Lamborghini Countach was faster, reading 0.852g, the same as the Ferrari 308GTS, and so was the 924's big-brother 930 Turbo (at 0.831). The standard 924 could do no better than a still-impressive 0.766g which should tell you just how finely tuned the 924S Turbo's suspension is.

Unfortunately, high blown turbo technology and suspension sorcery don't come cheaply and the pleasure of our 924S Turbo's company costs $25,800. Granted, we said the car is approaching the 911 in performance and handling. But that doesn't mean it has to cost almost as much. Still, price has not deterred people from buying the 911 and if the 924 Turbo ultimately becomes the air-cooled car's successor, cost won't count to those who demand the qualities Porsches have traditionally delivered.

PRICE

List price, all POE	$20,875
Price as tested	$25,800

Price as tested includes Sport Group ($2045), E-32 Option, inc air cond, removable roof panel, elect. adj outside mirrors ($1500), 2-tone paint ($895), AM/FM stereo cassette ($250), rear-window wiper ($235)

GENERAL

Curb weight, lb/kg	2850	1294
Test weight	3095	1405
Weight dist (with driver), f/r, %	50/50	
Wheelbase, in./mm	94.5	2400
Track, front/rear	55.9/54.8	1420/1392
Length	168.9	4290
Width	66.3	1684
Height	50.2	1275
Trunk space, cu ft/liters	10.4 + 7.9	295 + 224
Fuel capacity, U.S. gal./liters	16.4	62

ENGINE

Type		sohc inline 4
Bore x stroke, in./mm	3.41 x 3.32	86.5 x 84.4
Displacement, cu in./cc	121	1984
Compression ratio		7.5:1
Bhp @ rpm, SAE net/kW	143/107 @ 5500	
Torque @ rpm, lb-ft/Nm	147/199 @ 3000	
Fuel injection		Bosch K-Jetronic
Fuel requirement		unleaded, 91-oct

DRIVETRAIN

Transmission	5-sp manual
Gear ratios: 5th (0.60)	2.83:1
4th (0.93)	4.38:1
3rd (1.22)	5.75:1
2nd (1.78)	8.38:1
1st (3.16)	14.88:1
Final drive ratio	4.71:1

CHASSIS & BODY

Layout	front engine/rear drive
Body/frame	unit steel
Brake system	11.1-in. (283-mm) vented discs front, 11.4-in. (289-mm) vented discs rear; vacuum assisted
Wheels	forged alloy, 16 x 6J
Tires	Pirelli P7, 205/55VR-16
Steering type	rack & pinion
Turns, lock-to-lock	4.0
Suspension front/rear: MacPherson struts, lower A-arms, coil springs, tube shocks, anti-roll bar / semi-trailing arms, torsion bars, tube shocks, anti-roll bar	

CALCULATED DATA

Lb/bhp (test weight)	21.6
Mph/1000 rpm (5th gear)	27.3
Engine revs/mi (60 mph)	2200
R&T steering index	1.35
Brake swept area, sq in./ton	171

ROAD TEST RESULTS

ACCELERATION

Time to distance, sec:

0-100 ft	3.4
0-500 ft	9.2
0-1320 ft (¼ mi)	17.0
Speed at end of ¼ mi, mph	81.0

Time to speed, sec:

0-30 mph	2.9
0-50 mph	6.4
0-60 mph	9.3
0-70 mph	12.7
0-80 mph	16.7
0-100 mph	27.4

SPEEDS IN GEARS

5th gear (5000 rpm)	129
4th (6100)	103
3rd (6100)	80
2nd (6100)	54
1st (6100)	30

FUEL ECONOMY

Normal driving, mpg	20.0

BRAKES

Minimum stopping distances, ft:

From 60 mph	155
From 80 mph	258
Control in panic stop	very good
Pedal effort for 0.5g stop, lb	23

Fade: percent increase in pedal effort to maintain 0.5g deceleration in 6 stops from 60 mph.....nil

Overall brake rating	very good

HANDLING

Lateral accel, 100-ft radius, g	0.827
Speed thru 700-ft slalom, mph	61.5

INTERIOR NOISE

Constant 30 mph dBA	66
50 mph	69
70 mph	73

SPEEDOMETER ERROR

30 mph indicated is actually	29.5
60 mph	60.0

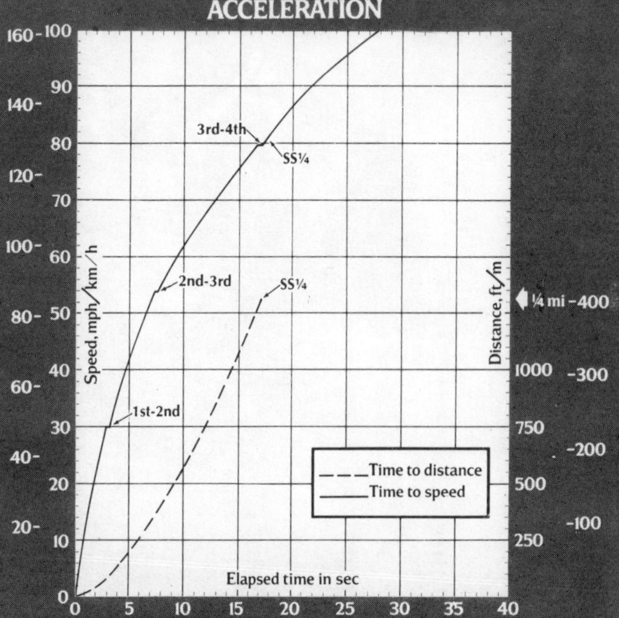

PORSCHE 917
VS
FERRARI 512

*Driving a pair of Le Mans legends
at Riverside*

BY SAM POSEY

PHOTOS BY JOHN LAMM

SOONER OR LATER somebody always asks me what was the fastest I ever drove. The answer is 235 mph. It happened 10 years ago in a specially bodied Ferrari 512S at Le Mans. It has always made great cocktail party talk, that 235 mph. What I don't always add is that even at that speed the Ferrari wasn't the fastest car at the race. The Porsche 917 did 245 mph.

Those two cars, the 512 and the 917, were the fastest cars in terms of top speed ever raced. In the Ferrari, the upshift from 4th to 5th was at 195 mph. At 210 mph the tach needle was still climbing swiftly, although by then there was no sense of acceleration. And there was no real feel in the steering either. The car reacted to air currents more than anything. The few times I was passed on the straight, the Ferrari pivoted on its axis like a

At the time, a unique twist in the rules required manufacturers to construct not just the usual five or six prototypes, but 25 each. This meant miniature assembly lines and the anomaly of pure racing cars lined up from wall to wall. It also meant numerous entries for the races. Le Mans had been targeted by Ferrari and Porsche as the key event and I'll never forget the sight of the Ferrari armada at the 1970 race: 11 512s, including four factory team cars. Porsche ran not only its own team but also hired John Wyer's, which had won the two previous Le Mans for Ford. The darker side to having so many prototypes around was that there weren't enough top drivers available to fill the seats, and rides often went to men who were out of their depth in a 600-bhp car. Englishman John Woolfe, driving a 917, was killed on the first

The Jo Siffert/Brian Redman Porsche 917 (#14) shares the front row of the pace lap at the 1970 Sebring 12-hour race with the Mario Andretti/Arturo Merzario Ferrari 512.

weather vane, as if the wheels weren't even in contact with the road. Unlike the other cars I have driven at Le Mans, it never reached terminal velocity, continuing instead to accelerate until it ran out of straightaway.

The opening laps of the race in 1970 were unforgettable. In qualifying, the 512s and the 917s had been equally spaced through the first six places. When the race started at 4:00 p.m., the track was dry but a storm was imminent. Rushing down Mulsanne at these unbelievable speeds, I could see the cars in front of me, weaving in oddly unpredictable ways, the sunlight glistening on their long tails. Beyond, the sky was pure black; lightning flickered constantly. When the rain began after about 5 laps, it came as a curtain of water stretched across the middle of the straight. I remember braking to 170 mph and easing the car into the rain as if entering a tunnel.

Without doubt, the two years these cars raced each other, 1970 and 1971, represented a high point in the history of long-distance racing; the two major manufacturers, each with racing in its blood, going at it with everything they had. Ferrari was determined to regain the prestige that Ford's dominant years had cost it; at the time it was designing the 512, Ferrari hadn't won a major endurance for three years, or Le Mans since 1965. For Porsche, the 917 was its first really big car and its first realistic chance to win Le Mans. Speeds were extraordinary. At Le Mans, the famous 7-liter MkIV Fords had lapped at 147 mph in 1967, a speed so imposing that the track organizers subsequently installed a chicane. But the very first 917 to hit the track in 1969 equaled the Ford's speed despite two liters less displacement and the chicane. The following year, Vic Elford raised the lap record to 150 mph and in 1971 Pedro Rodriguez' 917 did an incredible 155 mph.

lap in 1969. The car, he had admitted moments before the start, "frightens me to death."

Thanks to the fact I was regularly driving Can-Am and Formula 5000 cars, which had the edge on the 512s in cornering and acceleration, I found the Ferrari less daunting than it might have been otherwise. It did have the top speed but that was largely because it had the 3-mile long Mulsanne straight on which to run. What I worried about were the abnormal driving conditions you can encounter during a long-distance race. Fog. Rain at night. Enormous speed differentials between you and the slower cars.

Nevertheless, I loved the romance of driving in events which I had read about right from the moment I was first interested in racing. And the fact that I was able to contest them in the most legendary of all cars, the Ferrari, was more than just frosting on the cake. In the two years I drove 512s at Le Mans for Luigi Chinetti's North American Racing Team (NART) we finished 4th (in 1970) and 3rd (in 1971), on each occasion being the highest-placed Ferrari. Even then I knew it was foolishly romantic, but it meant a lot to me to be playing a role in Ferrari's history at Le Mans.

Not surprisingly, I viewed the Porsche 917, our nemesis, with a mixture of awe and dread. And since I never drove a 917 back then, I had always been curious about what it was really like. Now 10 years later *Road & Track* calls up and wonders if I'd be willing to drive a 917 and a 512 at Riverside.

The occasion was this:

Otis Chandler of the *Los Angeles Times* owns the Porsche and ⟫⟫→

rented Riverside for a day, inviting friends to bring their own cars and share the expense. This produced a stunning group of southern California's most prized and well preserved machines, including Chris Cord's 512S. Both Cord and Chandler had generously agreed that I could drive their cars.

It had been eight years since I had seen a 512 and a 917 together, but all at once there they were in the pit lane. I immediately realized that no matter how exciting either car is by itself, when they're together something special happens. I suppose part of this is just the triggering of memories of them wheel-to-wheel on the track, but there's something more, too, and perhaps this is because the cars are so visually different. Look at the Ferrari and you see a collection of details that invite close-up inspection. The air box. The flanges at the ends of the exhaust pipes. All the scoops and vents in the body. It's as if many different designers had worked on the car, each producing the most exquisite possible solution for a specific part, without any regard for how it would all ultimately fit together. The Porsche, by contrast, is all one piece, a single, unified, holistic piece of sculpture that you want to stand back from and take in all at once.

Chris Cord waved me over to his 512. Chris had the gifted mechanic Doanne Spencer prepare the car and it looked exactly

as it did 10 years ago. Doanne helped me in. It's an open car so you just step across the wide tank and you're standing on the seat, which, incidentally, is made of a distinctive weave of maroon upholstery which I've never seen anywhere but in a Ferrari. But then as you begin to slide in you notice your legs must bend awkwardly to the left because the pedals are almost on the centerline of the car. This was all right for the smaller drivers like Mario Andretti and Arturo Merzario but for Mike Parkes, Dan Gurney and me there was the feeling of being trapped from the waist down, and not supported at all around the shoulders.

Sitting in the cockpit I was reminded of how beautifully made the Ferrari is. Not only in the essentials, such as the castings for the hub carriers, but the finish and design of all the parts. Even details you wouldn't expect a company making a production run of 25 cars to bother with are exquisite. For example, the shifting mechanism looks as if it took a year of someone's life to design and build. The lever is the length of the palm of your hand and it's made from a substance called Micarta, a sort of laminated linen. The shift gate has been machined with the sharp clean edges you see in Bugattis.

There's a crispness to the starting procedure that any test pilot would enjoy. The initial twist of the key activates the high-pressure fuel pump; you watch the gauge as the needle flickers for an instant and then holds steady. At the second twist of the key the engine comes instantly to a fast idle at about 2000 rpm. There's no hesitancy whatsoever, no interval while the starter cranks the engine before it fires. It all comes on at once, idling with the metal-to-metal sounds of the cam gears and valves; you

don't hear the exhaust at all. The rules used to require that the engine be shut off during pit stops for refueling, and it was great not to have to worry about fumbling to restart the engine.

Getting the car moving was another matter. The clutch is very stiff, with a long throw, and only in the last half-inch of movement does it engage. Plus the accelerator linkage has an unusual amount of travel. The solution, of course, was just to dump the clutch and do a burn-out, and this was the approach favored by the factory team drivers, Mario in particular. But at NART we had instructions from Mr Chinetti to leave the pits "slowly, not crazy." I decided Chris and Doanne would prefer the Chinetti method and eased out of the pits and onto the track.

NART took the cars as they came from the factory, adjusted the anti-roll bars, shocks and spoilers to suit the track, but never the springs and basic alignment; those were left alone. The result was that all the 512s I drove had roughly the same feel, a

personality the way a road car has. So as I went into Riverside's turn 6 on that first lap I thought, "Now it's going to understeer, dammit" and there it was, that infuriating understeer that it always had in slow turns. It kept you from taking a natural line because you had to position the car for a very late apex, jockeying it around so that you could really jump on the throttle on the way out of the turn.

Under full acceleration the understeer vanished, the power poured through the rear wheels causing them to lose a bit of their lateral traction and the car became light, neutral. In the Chevrolet-engine Can-Am and F5000 cars there would be a surge of torque as you hit each new gear, then the power would flatten out. Formula 1 cars with Ford-Cosworth engines felt weak until about 1500 rpm from the redline, then they'd finally jerk you forward. But in the Ferrari, acceleration was a single even strain, a mighty and relentless rush. The magnificent gear lever allowed you to grab the shifts so fast and thoughtlessly that every trip down the straight was pure pleasure.

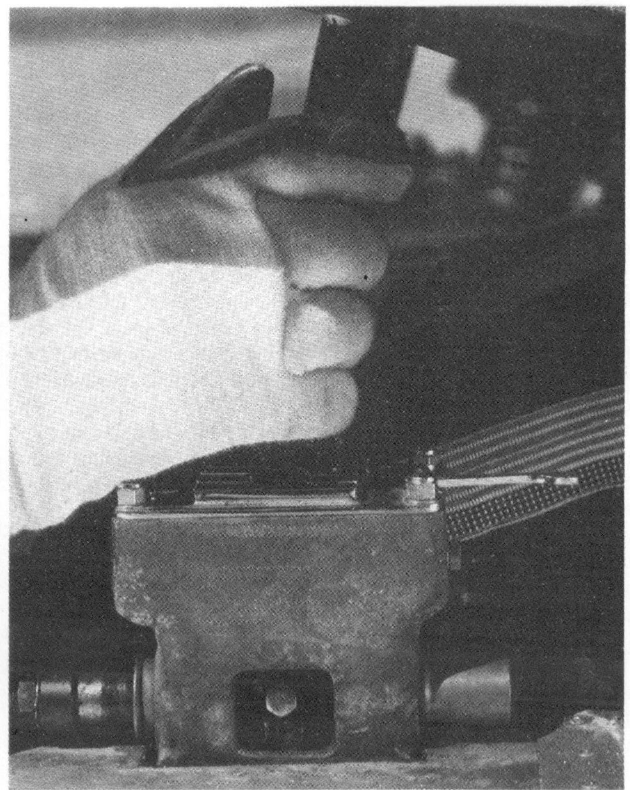

phenomenon of long-distance races that in the middle of the night drivers sometimes feel as if instead of traveling in circles they are making a vast journey away from the starting point. Of course, this sensation was heightened by the Ferrari's smooth acceleration, its flat cornering and above all by the never-ending roar of the engine, which eventually penetrated even the best earplugs and numbed your brain. At such times I used to feel an irrational affection for the Ferrari, as if I could trust it somehow, as if it would go on absorbing punishment and not let me down.

I had done 5, maybe 6 laps, when I noticed a twinge of pain in my right thigh. A couple more laps and the pain, which felt like a pulled muscle, was really sharp. All at once I remembered I always had that pain while driving the 512; it was caused by the cramped pedal area. I used to put in a lot of hours feeling my leg hurt that way. I stopped in the pits, limped a few feet away from the car and felt the pain fade, just as it always had. It was time to try the Porsche.

Specifically, it was a 917K built in 1969, a year before the Ferrari. It was a 4.5-liter, not the 5.0-liter that Porsche would use in 1970 and 1971. With 520 bhp, it gave away 80 bhp both to the Cord Ferrari and the later Porsches. Also, it had a 4-speed transmission instead of the 5-speed employed the following year.

The Ferrari's nicely machined shift mechanism (left). Below, Doanne Spencer talks with Sam about the interior of the Ferrari.

On paper, this early version of the 917 (chassis number 004) would be a fraction slower than the Ferrari.

Not that it looked that way. With its fearsome, sinister styling and the original pale blue and orange colors of the Gulf Oil/John Wyer team it appeared as invincible as ever. This was one of the short-tail cars, not the streamliners made specially for Le Mans. John Thomas, who looks after the 917 and the other Porsches in Otis Chandler's collection, told me that I could expect a top speed of about 190 mph. The Ferraris, incidentally, were all short-tail cars. The one exception was the quasi-streamliner Ronnie Bucknum and I drove in 1970, when I hit 235 mph. The Porsche streamliners were faster overall at Le Mans than their short-tail cars but our Ferrari, because of its extreme instability in fast turns, wasn't as good as the regular model.

Climbing aboard the 917, I discovered that despite its dramatic low silhouette and far forward driving position, it wasn't the least bit cramped inside. And the seat was raked at a comfortable angle and adjustable, whereas the Ferrari's was upright and a permanent part of the car. Ferrari teams were always stuck with pairing physically similar drivers together while the adjustability of the Porsche allowed them to put anybody in with anybody else, and if necessary install their ace in a car other than his own.

It was as if Porsche began the design of the whole car with the ⟫⟶

Then you got to a corner again and the car reverted to being a beast, a truck, something that made you so mad you hated the whole thing. The understeer! The heaviness of the steering! Usually if you've got understeer you've at least got light steering, but not in the 512. When I first drove one, I did a couple of laps and then came in. I was sure something was completely screwed up, but they said everything was fine. So I went to the Ferrari pits to see Mauro Forghieri, the engineer. At first he pretended not to know what was wrong but finally he admitted the kingpin inclination was off. (When the front suspension was revised for the 512M, the heavy steering, and most of the understeer, was eliminated completely. The M was a much faster car than the S, and my guess is that at least half the improvement came with this one change.)

Two laps went by. Three. I was remembering most of the old sensations. What I missed was the feeling I used to get only after many hours at the wheel. It's a rhythm that you come to when the alertness of the early hours gives way to reflexive motions and you drive like an automaton, guiding the car with an absolute economy of movement. The cockpit gets warm and womb-like, the instruments glow in the dark. It is a curious psychological driver, got him sitting in an ideal position right out in the middle of a large empty room and then bit by bit built the car around him. The Ferrari observed the traditional *look* of a cockpit whereas the Porsche designers provided a *space* that happened to be exactly the right size. For example, the area around the pedals was ample but to get my feet there they had to be maneuvered beneath the massive steering rack which crossed the forward part of the cockpit like a girder of a bridge. The designers had also plotted the driver's lines of sight. The instruments are buried in deep tunnels (to keep them from reflecting onto the windshield at night) and visibility to the rear consists of a view along a narrow trench in the back deck. With even the slightest movement of your head you become misaligned with these axes.

With such meticulous attention to the conceptual problems of cockpit design it was astonishing to find that no one seemed to have planned for the Porsche's steering wheel. Not only was it mounted at an awkward angle, tipped so that the top of the rim was almost out of reach and the bottom nearly in your chest, it was also so high in the car that the upper third of the rim was smack in the way of your view of the road. I was also surprised by the crude construction of the cockpit. Thick strips of black felt had been applied to the dashboard with a yellowish glue, much of which had oozed out and become matted in the felt. The roof and doors were rough, unsanded fiberglass painted dull gray. The gear lever resembled one from a VW, just a long shaft with a plastic ball at the top. And then there was the way the doors were hinged. They had to swing up to open; how would you get out if the car was upside down? Drivers of the 917 used to joke about this, saying they hoped the crash would be violent enough to break the car in two so they could step out of the back.

John Thomas had been leaning into the cockpit explaining the instruments, which included a light that warns you if the fan belt has failed—important information in an air-cooled car. When he shut the door I started the engine, just like that, the way you would in a street car, with none of the theater that had gone with starting the Ferrari.

My first big impression as I drove cautiously down the pit lane was of how low the car was; my eyes were level with people's knees. Accelerating onto the track, I noticed that the tops of the fenders were even with the horizon, which meant that the view out the front was strictly straight ahead, as if you were wearing blinders. Out the back of the car there was only that narrow, gunsight view, with the road not even visible until 200 ft behind the car. I once complained to Pedro Rodriguez that he had held me up for laps. "Sam," he had said, "I am sorry, but I can see

nothing out of the car." Now I was beginning to understand.

So there was this claustrophobic feeling but then, owing to the extreme forward placement of the seat and the fishbowl windshield, there was also a frighteningly immediate view of the road right in front of the car. The windshield extended down as far as my knees, and the onrushing road was visible to within a yard in front of my feet. With so much going on literally in my lap I had a peculiarly distorted impression of the rest of the car. I felt as though I were a great distance from the rear wheels. I felt like an astronaut at the tip of a rocket with seven stories of space vehicle somewhere below. In the esses I gave the apexes a wide berth, like the driver of a long truck who swings out to keep his rear wheels from hitting the curb.

I was intimidated, and yet I kept picking up signals from the car that said, hey, this is just a big VW. There was lots of body roll. The gear linkage seemed to be made of rubber. Exiting turn 6 I noticed how quiet the engine was. I found that as long as I avoided looking at the road directly in front of the car there was little feeling of acceleration, but I was fully aware of a sense of speed because of the car's disconcerting tendency to weave uncertainly from side to side, a result of suspension geometry that permitted the rear wheels to toe in and out.

I got around through turn 8 and stretching ahead was the piece of road this car had been made for: the straight. But instead of surging forward with the Ferrari's festival of upshifts and blaring exhaust, the Porsche alternately hummed and wheezed, pausing for a deep breath at each upshift while the linkage twisted its way to the far reaches of the car. Nevertheless, we were gathering momentum at a terrific rate and with the speed came an alarming loss of stability. The 917 acted as though it were on a high crowned road, first lurching off the crown to one side, then the other. We shot under the bridge and looming ahead was the kink in the straight, that little snick to the left at 180 mph, very similar to the *right*-hand kink in the Mulsanne, and I could see the trick was going to be to get the cycle of lurches to coincide with the bend. A glance in the rearview mirror revealed a token patch of roadway, a fifth of what I ought to see before veering diagonally across the track.

Only marginally controllable where the Ferrari had been like a rock, the 917 lunged through the kink and on down the short straight into turn 9. Still gaining speed! The Ferrari at this point had seemed to lose its acceleration but the Porsche was still going. Oh, I loved this! At Le Mans top speed is the essential weapon, the only winning hand for that 3-mile long rocket sled blast down the Mulsanne. Maybe it was just my imagination but Chandler's 917 even without the big engine or the streamlined body gave the impression of being able to cleave its way effortlessly through the air. Top speed! That's what these cars were good for! Suddenly I was remembering what it had been like at Le Mans, with me in the Ferrari and Jo Siffert or Elford or Rodriguez coming up from behind, their car jerking from side to side as if it were on trolley tracks. They'd start by and the guy would be so busy steering he couldn't even look over and wave. They'd be white behind those helmets but, dammit, they'd be *grinning!*

I continued for several laps. Cornering was easier than with the Ferrari because the Porsche wasn't afflicted with that grinding understeer. Driving it just as fast as it would go was no harder physically or mentally than driving at eight-tenths, a fact that was probably significant in the 24-hour races. Also, it was quiet and nearly vibration-free, other assets in the long run. I gradually came to think the rear wheels might not be so far back as I had first thought, although they never exactly caught up, either. As for the weaving and instability, once I realized the car wasn't going to plunge off the road I got used to it and relaxed. The shifting, however, remained incorrigible, the early part of the rush up any straight interrupted by the agonizing slow process of getting into the next gear.

All at once there were long shadows on the track; the day was over. People started to put the cars away.

Strange. At Le Mans dusk was just the beginning.

Letter from Europe

Porsche 924 Carrera GT & 1981 Porsche 911SC

W HILE IN Stuttgart I seized the opportunity to try a new 911SC and to use a 924 Carrera GT as personal transportation for other visits. I covered 600 miles in three days.

Though the U.S. and Japanese versions of the 911SC are essentially unchanged for 1981 (the main difference is a revision using an improved coil-spring clutch disc hub in place of the rubber-damped hub introduced two years ago), the European 911 engine has been restored to its full former glory. This occurred because of complaints from 911 fanatics that the 911SC was slower than the Carrera, which was discontinued in 1978. Improved performance has been achieved by a revision of the Carrera 3.0's valve timing and, above all, an increase of the compression ratio from 8.6 to 9.8:1. This makes the use of premium fuel imperative but improves the engine's specific fuel consumption and should make the current 911SC a more economical car. Compared with the Carrera 3.0, some power is wasted in a quieter exhaust system and in an air injection pump (both to meet stricter norms), but the gain of 16 bhp raises the power to 204 bhp DIN, 4 bhp more than the Carrera.

Porsche 924 Carrera GT.

Compared with my own 3.0, previous versions of the 911SC had less top-end urge but the latest edition has not only more of this but also improved throttle response in the lower-rpm ranges. This is probably a result of several detail changes to the K-Jetronic fuel injection system. Low-speed gearbox chatter has also been lessened. Modifications were made to the clutch and clutch disc to achieve this, correcting a fault that has been a 911 characteristic from its inception. I also found the gear change more positive than in my car and noted a general improvement in silence.

Time was too short for me to clock the latest 911. My impressions were correct, however, as evidenced by the test report recently published by *Auto, Motor und Sport* showing the car to be the fastest production 911 ever made. (The only exception would be a limited number of special lightweight Carrera RS models built in 1973–1974 essentially for competition purposes.) The car did 149.1 mph, 0–100 km/h (62.5 mph) in 5.9 sec, 0–160 km/h (99.4 mph) in 14.7 sec, and the standing-start kilometer in 25.9 sec. It also proved to be more economical than the older model.

These figures are virtually identical with those I obtained for the 924 Carrera GT. However, you have to pay 60,000 rather than 50,000 Deutsche Mark to purchase one, and you get more noise in the bargain, at least if you drive at 125 mph or more. Perhaps I should have written this in the past tense because only 400 Carrera GTs have been made to meet FISA homologation requirements in Group 4, and all of them have immediately found an owner. Additional models will be made only to rally or racing specifications.

The 924 Carrera GT is a special version of the production 924 Turbo. Its external differences include a conspicuous front airdam, widened front and rear fenders covering 16-in. diameter forged alloy wheels with 7- and 8-in. wide rims front and rear respectively, and an air intake on the hood. The springs and dampers are stiffer than in the normal 924 Turbo, and the car sits lower. A flush-fitting, bonded windscreen creates less aerodynamic disturbance while contributing to the stiffness of the body, and the widened fenders are made of a deformable polyurethane material capable of absorbing small bumps without damage. The air intake, bulging out on top of the hood, ducts cold air to a heat exchanger, cooling the compressed air before it reaches the cylinder head. This intercooler allows the engine's compression ratio to be increased from 7.5 to 8.5:1, improves the compressor efficiency and allows the boost to be raised from 9.3 to 10.7 psi. While the turbocharger's compressor remains unchanged, the turbine is of a different type. All this adds up to a power increase of 23 bhp to 210 bhp DIN at 6000 rpm, while excellent part-load efficiency is provided by a fully electronic ignition system with computer-controlled ignition timing.

The engine has acceptable power even in the lower ranges, but it progressively bursts into life as the turbocharger pressure rises. You can't actually watch it doing so, however, because there is no boost gauge. And from around 3500 rpm the engine just zooms up to its 6600-rpm limit. In normal driving with the five close-ratio gears and the engine's excellent mid-range torque (203 lb-ft at 3500 rpm), there is little point in pushing the revs higher than 5500 rpm. From a standing start, 60 mph is reached in 5.8 sec, 100 mph in 14.9 sec, and the kilometer mark is passed in a resounding 25.8 sec at more than 120 mph. All these times were achieved over an accurately measured base on a German *Autobahn* in freezing weather, when the anti-frost treatment of the roads reduced the grip and caused serious wheelspin problems when getting off the mark. Maximum-speed timings resulted in an average of a shade more than 149 mph.

Fuel consumption was another pleasant surprise. Excluding the performance tests, the Carrera registered 25.0 mpg (U.S.) when measured over more than 250 miles, two-thirds of which were on *Autobahnen,* cruising at 125 mph. At this speed there is little wind noise but the engine booms noticeably, a fault you don't have at the same speed in a 911.

The *true* speed (and consumption) must be noted. In contrast to most Porsches I have driven, this one had an optimistic speedometer and odometer. The official EEC consumption figures for this car are remarkable: Steady speed 90 km/h (55 mph), 120 km/h (75 mph), and the "urban cycle" figures average exactly the same as for the normally aspirated 924, a notably economical car.

Except for the stiff springing, which is rather excessive for a normal road-going car, this Porsche is a delight to drive. Among its virtues, I would put handling at the top of the list. With its nearly 50/50 weight distribution, the Pirelli P7-shod test car clung to the road impressively and was so stable and predictable that, as good as the engine is, handling must rank first. In this enumeration, I would include the beautifully positive, high-geared but never-too-heavy steering. The gear change also is a real delight despite the inertia of the driveshaft, which in Porsche's transaxle system (clutch in unit with the engine, gearbox in unit with the final drive) is part of the primary drive that has to be synchronized.

The brakes are basically the same. But as the wider wheels increase the track and the steering now has a positive scrub radius, the twin circuits are front/rear, rather than diagonal. Otherwise, the car varies little from the 924 Turbo. Thanks to careful attention to detail, even the low drag coefficient of 0.34 has been preserved in spite of the protruding fenders, though the overall drag is higher because of the larger frontal area.

Unfortunately, you can't buy a 924 Carrera Turbo anymore. But Porsche says there will be other special 924 turbocharged versions when development of the works' racing and rally cars has progressed further. ❧

THREE GRAN TURISMOS:

ALFA ROMEO GTV 6/2.5, DATSUN 280ZX TURBO & PORSCHE 924 TURBO

Our July 1976 Comparison Test revisited, with improved cars

BY HENRY N. MANNEY III

PHOTOS BY JOE RUSZ

FROM TIME TO time we get too many cars in the barn and have to do one of our famed multiple tests despite the fact that the whole process seems easier somehow when you do them one at a time. However, direct comparison does have its advantages, not the least of which is giving us a few days out of the house, as any Transylvanian car is going to get top marks tested solo from our resident Transylvanian car freak and so forth. An evaluation done by several members of the *gratin* may well produce several widely differing opinions but that's baseball, isn't it? and all to your advantage.

At any rate, the three Gran Turismos we gathered together here in the parking lot were, at first glance, superficially competitive in their specs but as it turned out, proved to appeal to completely different facets of the market. As we were not testing utilitarian little tin boxes but GTs, which theoretically at least reflect the best shots of certain manufacturers, this sort of buckshot engineering is inevitable and perhaps will save the reader from the sinking feeling that he has bought the wrong *bagnole* after all.

Much water has flowed under the bridge since sports car fever first hit this country; and we think that it is safe to say that any one of these machines, more or less prepared of course by the parent companies but certainly not race prepped (one of them just finished a 3500-mile trip across the country), could have given an extremely good account of itself in 1950 at Sebring or Pebble Beach, even with radio, air, and cigar lighter. Since the MG TC, competition-bred disc brakes, oil coolers, and the whole spectrum of suspension design on customer's cars have made tremendous strides. Another fairly recent spinoff from racing is the humble turbocharger, which gained some notoriety as a WWII quick fix for the problems of power fall-off at high altitudes and proved to be so efficient, compared to positive displacement Roots, vane or centrifugal blowers, that it shortly took over many racing duties.

Basically a turbocharger is a device in which the working or pumping end is powered more or less for free by a waterwheel in the exhaust flow which is going out willy-nilly anyway to pollute the atmosphere, curdle milk, frighten chickens and all those good things, thus giving you money for old rope. Actually it isn't that simple but an important advantage is that you get back a tad of punch from today's wheezing-clean engines, a side benefit being the potential for improved fuel mileage assuming you don't make a practice of sticking full boot into throttle at every opportunity. Makers of relatively gutless machinery have thus greeted this device with cries of joy, especially when said gutlessness was largely forced upon them by Gov't Ukase, and turbos have become part of the American Way of Life. Thus it is no surprise that two of the three GTs in our test have had turbos fitted in an effort to keep up with tomorrow, the third being an example of the classic approach: And don't we all know what there's no substitute for. It is fair to say that unlike some of the automobiles fitted with "old-fashioned" blowers we have driven, the two turbo cars, the Datsun and Porsche, displayed no evil habits and did the work for which they were designed; no water in the oil, no superheated jets of steam, no mice-et pistons, no cylinder heads looking like Armenian bread . . . isn't science wonderful? And the Alfa's added cubes ain't done it no harm neither.

Alfa Romeo GTV 6/2.5

ALFA ROMEO is definitely an engineers' company in spite of having been nationalized near the end of the Hitler War, I think. They have been making sporty cars since Garibaldi was a tot and have tried their hand at Formula 1 with considerable success, not to mention employing countless name drivers in the last 50 years. Equally superb engineers came from the Alfa school, not forgetting Enzo Ferrari of course, and even their most mundane postwar sedan had twincams, giant brakes and carefully thought-out suspension underneath, at a time when most of the industry was revamping 1925 stodges. Development proceeds at its own pace at Alfa—i.e., when the complete machine is ready—and the engineers think mostly of their own precious home market, which may explain why export customers get short shrift.

The angular little 1.3-liter Giulietta sedan grew into the pretty Sprint coupe and the light and very competitive Sprint Veloce; these evolved into the Giulias of various sizes (1.6 liters up) over something like 25 years and thus to the Alfetta and its cousins. The basic design was becoming a bit long in the tooth, especially as regards the engine, and eventually a nice V-6 engine sedan was rather slowly brought out. After endless fiddling about, Alfa decided that this particular sedan well might lay an egg in the

U.S. (they have never had too much luck with sedans over here, though the berline are often better) so decided to drop the nice V-6 into the Alfa GTV nee Alfetta GT they already had, a car that enjoyed the latest thing in suspension and styling but was lumbered with a somewhat noisy and antique 4-cylinder twincam bored out to the fins. *Voilà!* A transformation! Janus-like, Alfa at one stroke looked forward and backward; forward to a modern sports machine with smoothness, quiet, more than adequate power and exceptional fuel mileage . . . backward to the ghost of the sainted Tubolare, from which the GTV 6/2.5's manner of going derives.

No doubt most of you have some familiarity with the Alfetta GT clan with its front-mounted engine, rear-mounted transaxle and rakish lines. Very little has changed on the surface barring a tea tray on the hood but there are numerous mods . . . demon mods at that . . . lurking beneath the shiny surface. Most prominent of course is the lovely sohc V-6 engine (see data panels) which is not only as imposing as Alfa engines usually are but actually is distinguishable from the smog gear. How many times have you heard, "I wish the designers of this turkey had to work on it!"? The Alfa, in the manner of the Lancia Aurelia for example, has the clutch and all that jazz right aft so mechanical work is made relatively simple. For those sick of Spica injection, Alfas have now gone to Bosch and the ign is breakerless as well. To feed all this, a choice of several enlarged tanks is fitted which should cut down on luggage space in the trunk; "ours" was supposed to have a 20.4-gal. tank, but if so the trunk measured only 0.5 cu ft less (9.7 versus 10.2) and the car's fuel gauge didn't know about the larger tank. Probably 15–16 gal. is more like it.

Naturally enough with more power feeding through we were curious about what measures had been taken to keep everything nice underneath. The anti-roll bars and torsion bars have been made thicker, drivetrain ditto, driveshaft bigger diameter, a new double disc clutch which is a trifle sticky at times but didn't slip, bigger donuts, and also a new transaxle with more magnesium content in the reinforced casing (not just the usual Elektron) has been fitted, featuring a 4.10:1 final drive. At a guess, I would say that most of the above came out of the V-6 sedan if not all of it. The vented front discs are also bigger (10.5-in. dia) but the solid rear discs remain the same as before. Tires as well are upgraded as the GTV 6/2.5 is an honest 120-mph automobile if not more; Pirelli 195/60HR-15 tires in place of the old 185/60HR-14s clamped onto a slightly wider light-alloy wheel that is standard, not an option. It was difficult to find out if Alfa still kept its unique non-Ackermann steering in which the wheels are supposed to be parallel at all times but in any case a new hydraulic steering damper was fitted. In addition, turning circle is now tighter at 30.8 ft, curb weight a trifle more with bigger bits at 2840 lb vs the old 2660, and other dimensions down below were shifted about to accommodate the tires so that front track is now 54.0 in. vs the old 53.5, rear ditto 53.2 vs 53.4, wheelbase remains the same at 94.5 in. and the overall length has dropped a little to 167.7 vs 171.0 largely due to cleaning up the bumpers. Ready for the Mille Miglia?

Datsun 280ZX Turbo

WHEN FIRST introduced, the Datsun Z was sort of an undersized copy of an E-Type and none too clever at that. Since that time the industrious Japanese have assimilated more unfamiliar material than anybody in history and, assisted by vigorous advertising (about 70 percent of the U.S. population still doesn't know what an Alfa is), have sopped up a considerable part of the market. My view has been that the Z cars were sort of like 190SL Mercedes, the Schweizerdeutsch Thunderbird that rich folks bought for their girlfriends as it wasn't fast enough to get anyone hurt. No really serious person would own one. Not sporting. However, Datsun has since done a lot of racing and applied its findings to this 2.8-liter turbo-ized sohc six, about which you have already found a full road test in the May 1981 issue. Go forward three squares.

Porsche 924 Turbo

THE 924 is nothing new, really, having been brought out by the celebrated Porsche firm in 1976 (see comparison test, July 1976) as a mid-range variant of its originally VW-based series of sports cars. Times have changed, 22 thou is no longer regarded as mid-range but a *coup de fusil,* and in fact the 924 is more of an Audi GT than anything else, what with an engine in front cooled by water yet. However Audi does make good cars and with the DM being worth about twice as much relative to the dollar as it used to be, a German manufacturer cannot go on building specials and stay in the ballpark, especially when it makes three separate and distinctly different models with all sorts of options.

This turbo version is a successful effort to update the 924 range and incorporates quite a few changes for 1981, some of which you will receive with joy and others with the opposite, such as power windows being standard when lots of folks prefer the uncomplicated if stiff cranks (which, by the way, were fitted to our early production version). Other oddments at hand include disc brakes all around (was an S option), bhp is up to 154 at 5500, torque is up, fully electronic ign with idle stabilization and a new set of gear ratios. Porsche also changed the shift pattern around to a normal one, making the rather elevated 5th gear (0.73) more conventional in its position; whereas we all appreciated having 1st gear where it belongs on a non-racer, we also thought this was sort of expensive to meddle about redesigning gearboxes for linkages alone. But a Certain Happening at the dragstrip revealed that the gear-bag is in truth Audi, not Porsche as was the previous one. Will Audi gears be strong enough to take the turbo through endless Dairy Queen Grands Prix? Watch this space. Speaking of spaces, your manicurist couldn't wedge a nail file anywhere under the hood, said hood being the heaviest since that of an armored car we once tested. Whatever happened to light Porsches?

Route

AS WE seem to spend our lives in smoggy Orange County with its millions of stoplights and quota-filling cops, we thought that it would be nice to get out and act like GT people, ignoring the fact that most Alfas spend their time cruising for chicks, most Datsuns being washed and most Porsches in parking spaces marked "Doctor." Therefore we pointed our noses up the Pacific Coast and *les flonflon de* Carmel, staying as much as possible on State 1 and avoiding freeways in the interests of Evaluation even though most freeway surfaces in this part of the forest could give a cobblestone street points. Evaluation was right; the corniche road after San Luis Obispo features mountains on one side and a deep drop into the blue Pacific on the other and for spades it was simply hissing with rain. Our Boy Racer contingent was busy pressing on through all the sort of thing postmen are supposed to go through and don't, and all I could think of was the numerous times sections of Coast 1 have suddenly subsided to sea level.

Further edification was gained (we do this for you, Dear Reader) sharing time with a March GP machine and a rogue Can-Am at Laguna Seca race track. Bags of fun. Sure, these test cars aren't racers but what doesn't happen at 100 certainly won't happen at 50.

The trip homeward, sustained by an on-going and v nice lunch provided by Mrs Editor, was down the sporting Carmel Valley Rd (G 16), a dodge or two around Greenfield on G 15, a shot east toward the Diablo Range on G 13, down the old Coalinga Rd through a variety of fascinating valleys on what turned out to be the rift of the San Andreas Fault (!), and then a long haul down State 33 which parallels the freeway more or less through Blackwell's Corner (James Dean was killed here), a variety of dreary oil towns like Taft, and over a 5000 + ft mountain pass to Ventura and from thence back to base. Sometimes it rained, sometimes not, some roads were 2nd gear, some were flat enough to explore redlining the tach in top (so I'm told) and the cars got a good workout. Beats working in an office.

Alfa Romeo GTV 6/2.5.

Driving Impressions: Alfa Romeo GTV 6/2.5

AS FAR as handling was concerned, everyone except our Porsche nut gave the GTV full marks, the difference of opinion not being attributable so much to bias as to being used to a certain type of suspension. So many things in a car such as steering, handling, driving position and the like often reflect the accumulated experiences of a particular driver, be he short, tall. skinny or fat and if he is used to Volgas then he isn't going to feel comfortable in anything else. However everyone thought that the Alfa was a very forgiving automobile, not too picky about its line through a corner, with the traditional Italian propensity for mild understeer when a corner was overcooked. Suspension was very compliant while giving little body roll or sogginess, doing a super job of keeping all the wheels on the ground, and was the model of decorum as far as damping is concerned. To be sure there is some jiggliness at low speeds, visible more or less in all our cars, but this is characteristic of certain makes of tires (see any steel-belted radial) or of any well damped suspension. However you can either float like a Buick Dynaflow or go around corners; we drove down some pretty vile roads at speed and were quite comfortable and in control. In the same basket was the admirable steering, fed through a nice wood-rimmed wheel (one tester said that it got slippery) which drew a split vote no less, none of the negatives the same one, but disenchantment varied from "almost too light at speed" to "too much effort around slow corners" to "slight bump steering." The steering *is* light, accurate, direct, and typical Alfa but one wonders if the alignment problem (see Alfa story last month) has been really sorted. Brakes as well were rated quite good with much mmmm feel to tell you when the end was coming, so valuable in the wet.

Most complaints were about the bloody awful flexible rubbery shift linkage, which takes a fine touch to find the right slot, and its cousin, the new double disc clutch which not only is a trifle sticky (bringing a few horrendous dzimboums on selection from rest) but would chatter occasionally in reverse. Undoubtedly it will be the source of numerous warranty complaints but at least it never looked like slipping. Gear ratios themselves are well sorted, even if 1st is a trifle low for anything but Stelvio lacets, and as the assemblage is mounted way back there, is nice and quiet and doesn't blow superheated air up the shifter sock. Someone pulled the wooden knob off of the lever, a trick that happens occasionally at R&T, and one quick fix on Alfas is to saw a few inches off the shifter. Long levers are bad for the synchromesh anyway.

Driving positions taken generally also got a mixed reception, as one would guess from four testers of widely disproportionate sizes, but as it is infinitely adjustable via the seat and the tilting steering column . . .! I think the leather bucket seats are hard and prefer the Porsche ones, Winnie the Pooh thinks they are just right (but then he also thinks I'm bony), while yet another complained about too much lumbar support which the fourth tester loved. Pedals are not really in the best place for fast driving (the throttle is a bit high), view outwards and around were chosen

Datsun 280ZX Turbo.

Porsche 924 Turbo.

best by all even if it can get pretty hot in there with that high windscreen, and the traditional finish was To Taste. Alfa is usually good about instruments but in some positions of the tilting wheel, the speedo is masked (a large tach is dead center) and the larger ones are covered by a sheet of plain glass that does reflect in certain lights. Normal air circulation, heating and so forth was also judged Best although the Alfa doesn't have cool air to the feet for hot days; provided, however, are a battery of flap valves which send cooling or defrosting air practically everywhere else. The heater/defroster works nicely, is sufficiently sensitive to accept graduations, and even the egg nishning does its best. Finish is about 8 on a scale of 10 and yes you can get adults in the rear seats although kids have more room even if they do complain more too.

The engine is a joy, revving and pulling smoothly through its range and while perhaps at a tiny disadvantage compared with the turbo competitors in sheer power, especially with the slow change, shows absolutely no cam effect or temperament and furthermore is extremely quiet; in fact the whole car is. No more exhaust boom! The Alfa is geared just about right, and flexible besides being able to pull 5700 (the bottom of the stripey bit, 6200 the beginning of the orange region and the built-in rev limiter) if you give it a while, mostly in perfect smoothness

allowing for a bit of racket, we think, from the front tires. At this speed the front end becomes a bit light on a bumpy road but those loony enough to corner at 125 will undoubtedly fit a larger air dam. All temperatures remained normal at all times (it takes a lot to make the twin fans work) and it is a joy to see what a nice ⟩⟩⟩→

ENGINE & DRIVETRAIN

	Alfa Romeo GTV 6/2.5	Datsun 280ZX Turbo	Porsche 924 Turbo
Engine type	sohc V-6	sohc inline-6	sohc inline-4
Bore x stroke, mm	88.0 x 68.3	86.0 x 79.0	86.5 x 84.4
Displacement, cc	2492	2753	1984
Compression ratio	9.1:1	7.4:1	8.0:1
Bhp @ rpm, SAE net	154 @ 5500	180 @ 5600	154 @ 5500
Torque @ rpm, lb-ft	152 @ 3200	203 @ 2800	155 @ 3300
Fuel injection	Bosch L-Jetronic	Bosch L-Jetronic	Bosch K-Jetronic
Transmission	5-sp manual	3-sp automatic	5-sp manual
Final drive ratio	4.10:1	3.55:1	3.89:1
Engine speed @ 60 mph, rpm	2850	2980	2300

PERFORMANCE

	Alfa Romeo GTV 6/2.5	Datsun 280ZX Turbo	Porsche 924 Turbo
Acceleration:			
Time to distance, sec:			
0–1320 ft (¼ mi)	16.8	15.6	16.7
Speed at end of ¼ mi, mph	83.0	88.0	82.0
Time to speed, sec:			
0–30 mph	2.8	2.5	2.7
0–60 mph	9.1	7.4	9.2
0–80 mph	15.3	12.6	15.4
0–100 mph	25.2	22.0	26.6
Top speed, mph	125	129	127
Trip fuel economy, mpg	19.0	16.5	19.0
Brakes:			
Stopping distance, ft, from:			
60 mph	162	164	153
80 mph	293	288	269
Pedal effort for 0.5g stop, lb	20	20	23
Fade, % increase in effort,			
6 stops from 60 mph @ 0.5g	nil	25	nil
Overall brake rating	very good	very good	very good
Handling:			
Lateral acceleration, g	0.778	0.754	0.802
Speed through 700-ft slalom, mph	62.0	58.6	60.7
Interior noise, dBA:			
Idle in neutral	63	53	59
Maximum, 1st gear	83	82	83
Constant 30 mph	65	64	66
50 mph	69	68	69
70 mph	75	74	73

GENERAL DATA

	Alfa Romeo GTV 6/2.5	Datsun 280ZX Turbo	Porsche 924 Turbo
Base price	$16,983	est $16,500	$21,500
Price as tested[1]	est $17,433	est $16,660	est $23,125
Curb weight, lb	2840	2995	2850
Test weight	3030	3090	3050
Weight distribution (with driver), f/r, %	50/50	51/49	50/50
Wheelbase, in.	94.5	91.3	94.5
Track, f/r	54.0/53.2	54.9/54.7	55.9/54.8
Length	167.7	174.0	168.9
Width	65.5	66.5	66.3
Height	52.4	51.0	50.2
Fuel capacity, U.S. gal.	20.4	21.1	16.4
Brake system, f/r	disc/disc	disc/disc	disc/disc
Wheel size	15 x 6J	15 x 6JJ	15 x 6J
Tires	Pirelli P6, 195/60HR-15	Bridgestone Potenza, P205/60R-15	Pirelli CN36, 185/70VR-15
Suspension, f/r	ind tor/ De Dion coil	ind coil/ ind coil	ind coil/ ind tor

[1]Price as tested includes: For the Alfa Romeo GTV 6/2.5, std equip (air cond, elect. window lifts, leather interior, alloy wheels), dlr installed AM/FM stereo (est $450); for the Datsun 280ZX Turbo, std equip (air cond, auto trans, cruise control, elect. window lifts, T-bar roof, alloy wheels, digital AM/FM stereo/cassette), 2-tone paint ($160); for the Porsche 924 Turbo, std equip (air cond, elect. window lifts, alloy wheels), sunroof ($510), AM/FM stereo/cassette (est $450), metallic paint ($445), elect. adj outside mirrors ($220).

safe car Alfa puts out. No twist, no wiggle, no rattle, no bad habits. Real quality.

Driving Impressions: Datsun 280ZX Turbo

UPON BEING assigned to drive this for a spell with its two-tone finish, automatic transmission, and general air of a Cad's Car, I said o no why me, especially with a wet and rainy Coast One coming up. Was never so surprised in all my life. Driving position was first-class, seats agreeable even if lacking a little support laterally, every instrument in the world was in plain view, the softened-up but under-damped suspension was comfortable, and the automatic worked perfectly even when using it for a brake, and the turbo-ized engine pulled and pulled and pulled with the right sort of engine note and no naughty noises or habits. Gad. No cam effect, the oil stayed clean till the end unlike the Porsche, and a few fast trips around Laguna Seca proved that while the steering may feel distinctly self-centering, the handling isn't bad with basically mild oversteer in extremis after it goes clunk over on the limit stops (you have to chuck it about like a stocker); you aren't going to get hurt unless by going out on a limb a lot further than we did.

Fundamentally the Datsun has less "soul," if that matters, than the pursang Alfa and Porsche and also what felt like about 800 lb more weight (actually about 150), as the rather insensitive brakes were difficult in the wet and also faded from time to time; perhaps a little red brake light that occasionally lit was for reminders. Harder pads might help. Actually the Japanese model is very stable if a bit clumpy at times and would probably profit by replacement of the off-road type Bridgestones which not only offer what seems like a lot of unsprung weight but seem too hard a mix, like most Oriental tires, to do much good in the wet. But then they probably wear forever. I liked the Datsun, in the face of some opposition. Gentlemen, meet the world's best boulevard sports car, one that can just about hold its end up with the real racers. Do you suppose that the local junkyard has a suitable transmission?

Driving Impressions: Porsche 924 Turbo

OLD PORSCHE fans will like the 924 Turbo as far as excellent roadholding, accurate steering, rigid carcass, tanklike visibility, firm seats, short throws to the gearbox, and well snubbed suspension are concerned; there is even a hint of the *wischen* action beloved of the older swing-axle coupes even though this car is laid out differently. Brakes are also first-class if, perhaps, with a little less feel than the Alfa's, and again this is a matter of personal preference as the two are as different as chalk from cheese. Many complaints were registered about the small and low-set steering wheel that made getting in a bother and, for some drivers, rubbed against the thighs. Nobody on OUR mailing list likes being rubbed with brown plastic, of which there is plenty about. Thanks to the comfy seats which surprisingly affected a couple of us after 100 miles as the padded roll tapers at

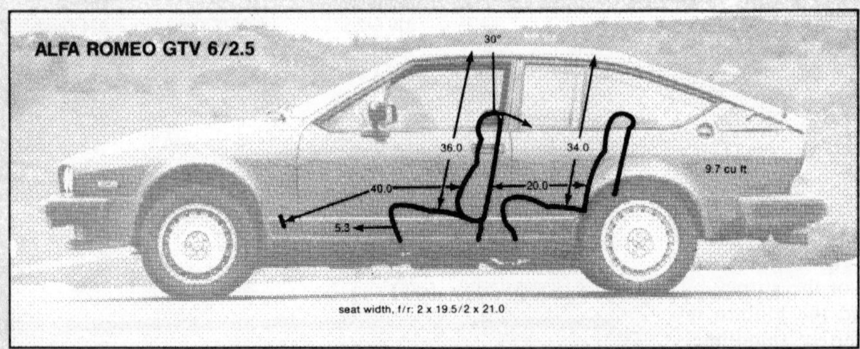

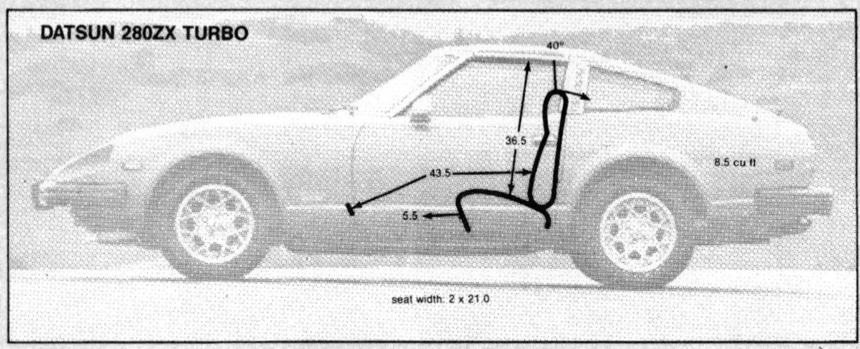

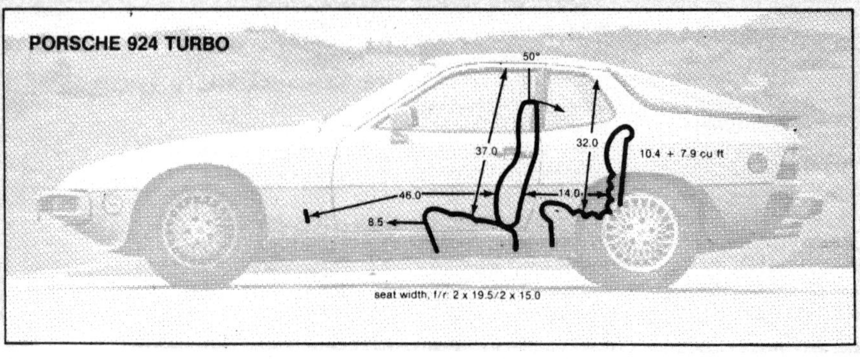

TECHNICAL INTRIGUES

THERE ARE FEW limits to the lengths we will go to entertain, edify, inform and otherwise astound you, our faithful reader (not to mention the potential gains to be had from winning bar bets). And after generating all the various performance figures for the Alfa GTV 6/2.5, the Datsun 280ZX Turbo and the Porsche 924 Turbo, we discovered some interesting parallels to a comparison test that appeared in *Road & Track* exactly five years ago.

The 924 was brand new at the time, and we journeyed up to Monterey, California for its introduction and to see how it stacked up against two of its closest competitors. Would you believe the Alfetta GT and the 280Z? There are times when one number is worth 1000 words and this is one of them. You'll note that the 924 we tested in 1976 was 0.1 second quicker from 0–60 mph and in the quarter mile than the Alfetta GT. The same is true of the 924 Turbo versus the GTV 6/2.5. The 280Z was about 2.5 sec quicker to 60 mph and around 1.0 sec quicker in the quarter mile than its two rivals then and a similar margin exists today. Incredible, isn't it?

You'll find other fascinating similarities in top speed and in our objective handling tests: Make note of the impressive gains in slalom and skidpad per-the top, the driver is well secured in front of the speedo (which one of us couldn't see v well) and an upside-down mounted tach which we all could. Visibility of these dials was not all it should be as they are very deep-set and in dull colors at that; too many idiot lights abound and in fact several of the useful gauges like the oil pressure and clock are tucked off tiny-wise in the console. Nobody could get much joy out of the heater/fresh air arrangement in spite of the usual whispery fan, defroster included. The 924 Turbo did have a sunroof, anyway, v nice for pottering around on sunny days and on the cross side, someone commented about the wipers lifting off at speed. None of the cars have opening wind wings and if the window is cracked, quite a few drops fall in from the reverse tumblehome of the roof; the Datsun spits rain in from the front slot and the Alfa is marginally the best, what with its fixed versions of what used to pivot open.

Porsche suspension is a marvelous device and perhaps had the "best feet" in ballet parlance at speed, being the most neutral of all three, although an awful lot of tire noise seems to be transmitted into the body as if everything were a bit under-gauge . . . not true! The Alfa is the lightest of the three cars by 10 lb. The Porsche also felt as if the wheelbase were much shorter than the others, purely a matter of habitude I suppose. Most grumbles however came about the engine, which really did have a pronounced blower effect midway up the range but not enough to bother anyone in the wet. At times it is sweet . . . well, sweet for an Audi which tends to feel as if the block castings were ⅛ in.

thick . . . but rough as last year's corncob when pulling hard under the collar. Furthermore the exaggerated gap between 4th and 5th on the touring Audi box (possibly fitted to give better fleet mpg figures) means that one can tick along lazily at no rpm doing 55–60 but due to the engine's design along with the blower, means that a vacant powerless hole is often encountered at mid revs when the loud pedal is pressed. Thus a change-down is necessary, rather a drag with all the extra rah unless the Baron von Hanstein act is onstage. Curiously enough, the engine mit turbo was completely docile at 50-ish speeds/gentle throttle, though you'll get a decidedly split opinion if you ask us what we think of its low-end torque. Certainly the performance is better, possibly in view of club racing, and whether the buyer wants to put up with the 924's other peculiarities is his business.

Charts

POINT RATINGS. At the beginning of these tests, the Tech Ed hands out a boxed-off bit of paper divided under headings of Engine, Brakes, Ride, Finish and so forth in order to rate the cars, further confusion being caused by orders to give each category a rating from 1 (the pits) to 10 (perfection). These at least ensure that we pay attention, even if the chap who does the report pays little or no count to them. However the ratings do reveal some mad flights of fancy which can be stored up for later use as Blackmail. Two staffers gave no 10s, one only one (Datsun instrumentation) and the other, three 10s, one of which was again ⟶

formance for all three cars. And, proving that quick cars don't necessarily have to be noisy, all three are considerably quieter than their 5-year-old counterparts.

A brief disclaimer about fuel economy: Today it's lower for all three models and certainly no surprise when you consider that one (the Alfa) has a larger engine, the other two are now turbocharged and our driving styles resulted in a heavy foot on each car's throttle for much of the trip. For more normal driving, I'd bet each car will return fuel economy as good or better than its predecessor.

The only anomaly for which we have no explanation is the longer panic braking distances for the GTV 6/2.5 and the ZX Turbo.

Price is certainly a subjective and controversial area, but if you do the sums, you'll find that today's Alfa GTV 6/2.5, 280ZX Turbo and 924 Turbo cost approximately 2.0, 2.5 and 2.3 times as much (comparing base prices) as their predecessors. Obviously the weaker position of the U.S. dollar relative to other currencies plays a part in the increases, as does inflation. And don't forget that each of these cars contains appreciably more standard equipment. All in all, we'd say that every one of these cars offers value for the dollar—the Alfa and Datsun more

so, as we discovered in our balloting. And without a doubt the answer to the oft

asked question "Are today's cars better?" is a resounding yes.—*John Dinkel*

PERFORMANCE*			
	Alfa Romeo Alfetta GT	Datsun 280Z	Porsche 924
Acceleration:			
Time to distance, sec:			
0–1320 ft (¼ mi)	18.4	17.3	18.3
Speed at end of ¼ mi, mph	75.5	81.0	75.0
Time to speed, sec:			
0–30 mph	4.4	3.2	3.8
0–60 mph	12.0	9.4	11.9
0–90 mph	27.8	22.1	29.1
Top speed, mph	112	119	111
Trip fuel economy, mpg	23.5	21.5	24.5
Brakes:			
Stopping distance, ft, from:			
60 mph	152	130	153
80 mph	252	221	294
Pedal effort for 0.5g stop, lb	14	35	25
Fade, % increase in effort,			
6 stops from 60 mph @ 0.5g	43	nil	nil
Overall brake rating	very good	excellent	very good
Handling:			
Lateral acceleration, g	0.726	0.720	0.766
Speed through 700-ft slalom, mph	54.5	53.9	58.9
Interior noise, dBA:			
Idle in neutral	62	54	65
Maximum, 1st gear	84	85	83
Constant 30 mph	71	68	71
50 mph	78	71	73
70 mph	75	75	77

*From July 1976 Comparison Test.

for Datsun instrumentation and the others for Alfa vision and quietness. Naturally the scale of gradations differed with two of us going no lower than 6 and a lone dissident dropping to 3 counts, both of them on the Porsche for ventilation and driving position, although the latter judgment is a bit harsh! Some of the comments such as "feel of many pieces," "eccentric wheel only exacerbates the problem," "terrible," "you will die in here on a sunny day," "Corvettish," "looks like a realtor's den," "typically kraut," "GM should take a lesson," "makes my back ake," "knee bender," and "wouldn't have it as a gift," might possibly make the engineer's ears burn but finding these things out on the road instead of in the showroom with the salesman kissing your wingtips might just be useful.

The first six categories of Engine, Gearbox, Steering, Brakes, Ride and Handling are of course both objective and subjective but do tend to get to the meat, so to speak. In these categories the Alfa registered 195 pts to the Datsun's 161 to the Porsche's 182, indicative to be sure but subject to road conditions where driven as well as a bit of old-fashioned prejudice. We are human too although that might be arguable in certain quarters. At any rate, bearing in mind that the rear seats or absence of were left out of the reckoning, the final results were by the four drivers (we also had a pix and picnic car along) in our order—

	Alfa	Datsun	Porsche
A.	154	124	140
B.	164	160	159
C.	136	145	144 (guess who)
D.	181	165	149 (guess who)

which is perhaps as unanimous as we ever get. All these added up to Alfa 635, Datsun 594, and Porsche 592 which may give Porsche something to think about and Datsun as well. Certainly the Alfa has its vices, as do they all, but the score gives a good indication that an entertaining, well balanced automobile is still worth the money. The next move is up to Alfa and its often maligned dealer network.

As an addendum, we were asked to rate the cars under personal choice "Forget Price" and personal choice "Consider Price," not an idle pursuit in these days of inflation. The results were a little surprising; two of us opted for the finishing order (alphabetical!) in both categories, one tester did 1-3-2 and then 1-2-3 fairly normal, while the last staffer screwed up all the perms with 2-3-1 and then 1-3-2, showing at least that 6 or 7 thou does make a difference in the real world. Fascinating, innit?

EDITORS' CHOICES*

Alfa Romeo GTV 6/2.5	1st/1st	1st/1st	1st/2nd	1st/1st
Datsun 280ZX Turbo	2nd/3rd	2nd/2nd	3rd/3rd	2nd/2nd
Porsche 924 Turbo	3rd/2nd	3rd/3rd	2nd/1st	3rd/3rd

*Each first ordering considers price/second is price-independent.

CUMULATIVE RATINGS—SUBJECTIVE EVALUATIONS

	Alfa Romeo GTV 6/2.5	Datsun 280ZX Turbo	Porsche 924 Turbo
Performance:			
Engine	32	**34**	23
Gearbox	**31**	30	27
Steering	**33**	23	**33**
Brakes	32	24	**35**
Ride	**31**	27	30
Handling	**36**	23	34
Body structure	35	30	**36**
Subtotals	230	191	218
Comfort/Controls:			
Driving position	30	**33**	26
Controls	**33**	32	29
Instrumentation	27	**39**	32
Outward vision	**32**	**32**	29
Quietness	32	**34**	29
Heat/vent/air conditioning	**35**	28	20
Ingress/egress	**33**	31	22
Front seat	31	**32**	**32**
Luggage & loading	**30**	**30**	**30**
Subtotals	283	291	249
Design/Styling:			
Exterior styling	**32**	25	30
Exterior finish	30	29	**33**
Interior styling	**29**	28	**29**
Interior finish	31	30	**33**
Subtotals	122	112	125
Totals	635	594	592

PORSCHE 924
WEISSACH

A *"commemorative edition" of the lowest-priced Porsche*

BEHIND THE 1976 introduction of the Porsche 924 was a remarkable story of corporate vicissitudes. The story began in 1970, soon after the Porsche 914 (the mid-engine 2-seater) had been put on the market. Though the 914 was sold in America as a Porsche, in Europe it was called a VW-Porsche, a name appropriate to both its engineering content (Porsche chassis and body, Volkswagen engine in the 4-cylinder version) and marketing organization (the VW-Porsche Vertriebs-gesellschaft, a joint company founded to handle the car in Europe). The Vertriebsgesellschaft wanted to get right to work on a successor and gave Porsche's Engineering Center the contract to develop one.

What this all meant was that Porsche, working under contract to the joint VW-Porsche company, developed a car that was intended to be sold under the name VW-Porsche. But those were troubled times at VW. The first big management shakeup of the period after the project got underway came when Rudolf Leiding took over the directorship of VW, killed the VW-Porsche organization and adopted the coming new sports car into the VW-Audi model lineup, with Porsche continuing the engineering and design work on it. But before Leiding could put all his ambitious new-model plans into production, he left in a huff and the present director, Toni Schmücker, took over.

Schmücker, however, did not see the 2+2 coupe as fitting into the VW or Audi lines and decided to abandon the project. Porsche, in a way left holding the bag and not wanting to be without a successor to the 914, then bought the project back from ➤➤➤

AT A GLANCE

	Porsche 924 Weissach	Alfa Romeo GTV 6/2.5	Mazda RX7 GSL
Curb weight, lb	2650	2840	2455
Engine	inline-4	V-6	2-rotor Wankel
Transmission	5-sp M	5-sp M	5-sp M
0–60 mph, sec	10.6	9.1	9.7
Standing ¼ mi, sec	17.6	16.8	17.1
Speed at end of ¼ mi, mph	78.0	83.0	80.5
Stopping distance from 60 mph, ft	150	162	170
Interior noise at 50 mph, dBA	70	69	71
Lateral acceleration, g	0.774	0.778	0.767
Slalom speed, mph	60.0	62.0	58.6
Fuel economy, mpg	22.5	19.0*	21.0

* Trip fuel economy

VW just before the car's planned introduction, and went ahead with production plans pretty much as VW would have—that is, with the car built at the Audi plant in Neckarsulm, West Germany rather than at Porsche's plant in Stuttgart. To identify the first production models as Porsches, a block-lettered decal was put onto the car's rear window, and it was rumored widely at the time that Porsche immediately set the price about 50 percent higher than VW had intended it to be.

The earliest 924s were not exactly what one would expect to be carrying the Porsche name, which considering the circumstances was no surprise. There was no missing the heritage of its engine—Audi—because like the other 4-cylinder engines of that firm, it was rough and noisy. The car itself was plenty noisy too, with a lot of wind noise and intense road rumble on poorly surfaced pavement. Despite the source of the 924's suspension, though—at the front MacPherson struts from the VW Super Beetle and lower A-arms from the Rabbit, at the rear semi-trailing arms from the Beetle—Porsche had made sure the car handled right. If not a "real Porsche," at least it showed signs of having the potential to become one.

That is essentially what has been happening to the Porsche 924 ever since 1976. Barely a year after its introduction, an upgraded 924 was introduced, with improved ride and noise control and, for the North American market, a boost in power from the original 95 bhp to the present 110. Running changes of a minor nature were phased in right along, but the next larger improvements came with the 1980 model, when the U.S. version got new emission control via a 3-way catalyst and oxygen sensor, an improved sound-insulation package, some handsome new interior fabrics and further ride improvements. Meanwhile, the 924 Turbo appeared, and since then Porsche has been busy cooking up special versions of the regular 924 to boost its sales—which have taken a beating at the hands of the much lower-priced, but

generally comparable, Mazda RX-7. All the same, more than 100,000 924s have been built so far.

Which brings us up to 1981 and the 924 Weissach. There couldn't be a more appropriate name for a "commemorative edition" 924—indeed any 924, since Weissach (pronounced approximately *vie-sock*) is the little town near which Porsche's Engineering Center is located. Painted a metallic color Porsche dubs Platinum, the Weissach is also a limited edition: just 400 were built, and there may be none left by the time this is printed. (Most of its features, however, are available separately.) It carries spoked alloy wheels like those our first test 924 Turbo had, in 15-in. size with Pirelli P6 tires size 205/60R-15. Like the Turbo, it has a rubber spoiler nicely set onto the huge rear window/hatch; ⟶

Comfy but durable-looking: the Weissach's brown-and-beige tweed interior. Pull-back cover for cargo compartment is a valuable feature.

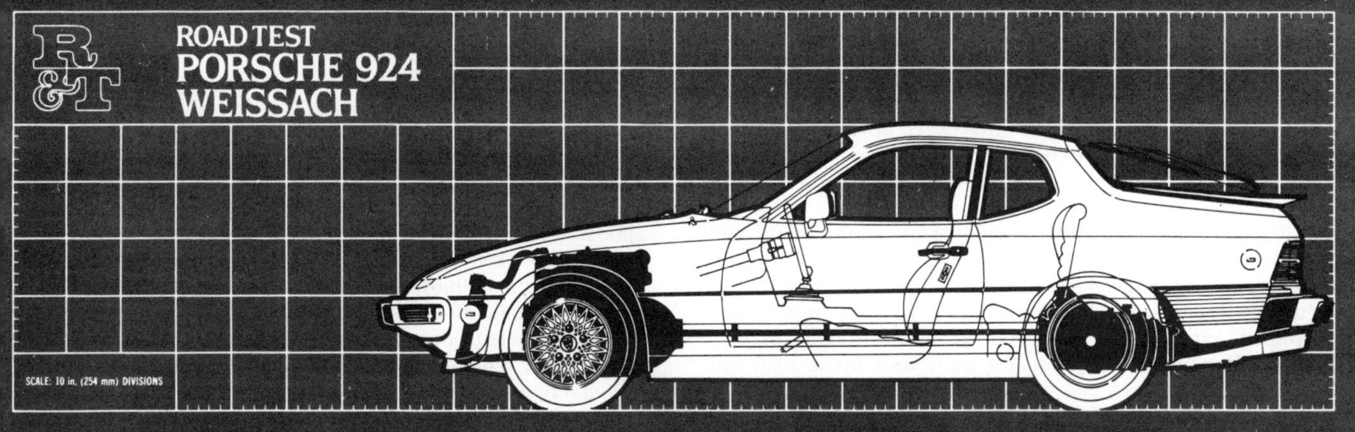

PRICE

List price, all POE $19,500
Price as tested $20,470
Price as tested includes std equip (air cond, sunroof, elect. window lifts, alloy wheels & Pirelli P6 tires), AM/FM stereo/cassette ($795), cassette holder ($35)

IMPORTER

Porsche-Audi Div, 818 Sylvan Ave, Englewood Cliffs, N.J. 07632

GENERAL

Curb weight, lb/kg	2650	1205
Test weight	2805	1272
Weight dist (with driver), f/r, %		48/52
Wheelbase, in./mm	94.5	2400
Track, front/rear	55.8/54.8	1418/1392
Length	170.1	4321
Width	66.3	1684
Height	50.0	1270
Ground clearance	5.9	150
Overhang, f/r	34.4/41.2	874/1046
Trunk space, cu ft/liters	10.4 + 7.9	295 + 224
Fuel capacity, U.S. gal./liters	16.4	62

CALCULATED DATA

Lb/bhp (test weight)	25.5
Mph/1000 rpm (5th gear)	25.0
Engine revs/mi (60 mph)	2400
Piston travel, ft/mi	1330
R&T steering index	1.21
Brake swept area, sq in./ton	192

ENGINE

Type		sohc inline-4
Bore x stroke, in./mm	3.41 x 3.32	86.5 x 84.4
Displacement, cu in./cc	121	1984
Compression ratio		9.0:1
Bhp @ rpm, SAE net/kW	110/82 @ 5750	
Equivalent mph / km/h		144/232
Torque @ rpm, lb-ft/Nm	111/150 @ 3500	
Equivalent mph / km/h		87/162
Fuel injection		Bosch K-Jetronic
Fuel requirement		unleaded, 91-oct
Exhaust-emission control equipment: 3-way catalyst		

DRIVETRAIN

Transmission		5-sp manual
Gear ratios: 5th (0.73)		3.00:1
4th (0.97)		3.99:1
3rd (1.36)		5.59:1
2nd (2.13)		8.75:1
1st (3.60)		14.80:1
Final drive ratio		4.11:1

MAINTENANCE

Service intervals, mi:

Oil/filter change	7500/7500
Chassis lube	none
Tuneup	15,000
Warranty, mo/mi	12/unlimited

ACCOMMODATION

Seating capacity, persons		2+2
Head room, f/r, in./mm	37.0/32.5	940/825
Seat width, f/r .2 x 19.5/2 x 15.0		.2 x 495/2 x 381
Seatback adjustment, deg		50

CHASSIS & BODY

Layout		front engine/rear drive
Body/frame		unit steel
Brake system ....10.1-in. (257-mm) discs front, 9.1 x 1.5-in. (231 x 38-mm) drums rear; vacuum assisted		
Swept area, sq in./sq cm	269	1736
Wheels		cast alloy, 15 x 6J
Tires		Pirelli P6; 205/60R-15
Steering type		rack & pinion
Overall ratio		19.2:1
Turns, lock-to-lock		4.0
Turning circle, ft/m	30.3	9.2

Front suspension: MacPherson struts, lower A-arms, coil springs, tube shocks, anti-roll bar

Rear suspension: semi-trailing arms, torsion bars, tube shocks, anti-roll bar

INSTRUMENTATION

Instruments: 85-mph speedo, 6600-rpm tach, 99,999 odo, 999.9 trip odo, oil press., coolant temp, fuel level, clock

Warning lights: oil press., brake system, handbrake, alternator, low fuel, oxygen sensor, rear-window heat, seatbelts, hazard, high beam, directionals

RELIABILITY

Owners of earlier-model Porsches reported 11 problem areas and 4 disabling reliability areas compared to overall Owner Survey averages of 11/6. So we expect the overall reliability of the 924 to be average.

ROAD TEST RESULTS

ACCELERATION

Time to distance, sec:

0-100 ft	3.5
0-500 ft	9.4
0-1320 ft (¼ mi)	17.6
Speed at end of ¼ mi, mph	78.0

Time to speed, sec:

0-30 mph	3.3
0-60 mph	10.6
0-90 mph	25.0

SPEEDS IN GEARS

5th gear (4900 rpm)	123
4th (6600)	123
3rd (6600)	84
2nd (6600)	54
1st (6600)	32

FUEL ECONOMY

Normal driving, mpg 22.5

HANDLING

Lateral accel, 100-ft radius, g ...0.774
Speed thru 700-ft slalom, mph ...60.0

BRAKES

Minimum stopping distances, ft:

From 60 mph	150
From 80 mph	280
Control in panic stop	very good
Pedal effort for 0.5g stop, lb	25

Fade: percent increase in pedal effort to maintain 0.5g deceleration in 6 stops from 60 mph ....nil
Overall brake rating ....very good

INTERIOR NOISE

Idle in neutral, dBA	58
Maximum, 1st gear	88
Constant 30 mph	67
50 mph	70
70 mph	76
90 mph	81

SPEEDOMETER ERROR

30 mph indicated is actually	30.0
60 mph	60.5
80 mph	81.0

ACCELERATION

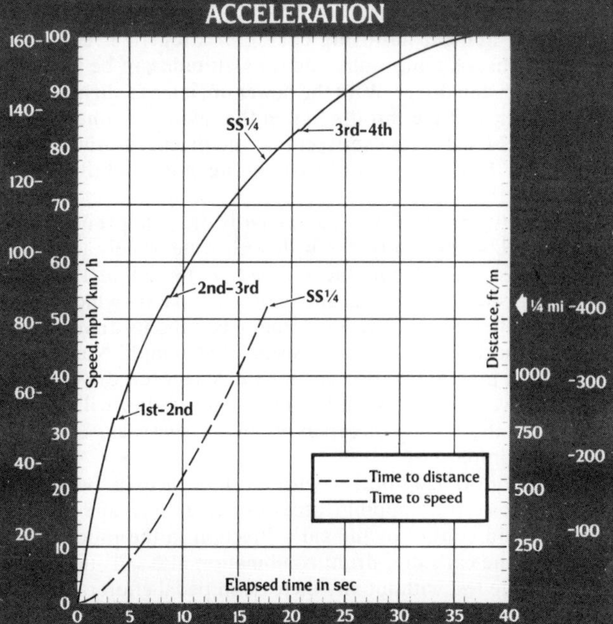

inside there are electric window lifts, tweed upholstery, leather steering-wheel rim and shift boot, and air conditioning. A takeoff sunroof and remote-controlled electric mirrors are also parts of the package.

All this comes to $19,500, including air conditioning and electric window lifts. With this price the 924 has now reached just about double its original price, which at that time seemed high. In the interim it has also become a much better, and much better-equipped, car. But it still has that relatively rough, not very powerful 4-cylinder engine: does it really stack up as a $20,000 automobile?

The answer to that question, in the final analysis, must be a personal one. But one thing is sure: the 924 is a far better car than it was when first offered for sale in 1976. Its trump card remains

handling, along with other strengths in areas where the transmission of forces through tires to the pavement is concerned. The 924 is an extremely well balanced car, as it well should be with its engine at the front, its transmission at the rear and a slight weight bias toward the rear.

With its masses so distributed, the 924 has less resistance to changing directions than a lot of cars; translated into more practical terms this means quick response to the steering and basically neutral handling characteristics. In keeping with Porsche chassis philosophy, the car's attitude can be influenced strongly by the driver. With the power on, it tends slightly toward understeer. Lift the throttle foot in the middle of a hard corner, and the 924 goes into oversteer that the practiced driver can use skillfully but which could catch the not-so-skilled driver unawares.

In reality, though, it takes a foolhardy driver to get into trouble with the 924, especially one with such wide wheels and tires as this Weissach had. On the skidpad, it turned in the strictly upperclass value of 0.774g, and in the slalom test—which requires a quite different set of reflexes from a car and its driver—it went through at the very creditable speed of 60.0 mph. Numbers like this mean high limits for a car; we think very few drivers indeed will ever venture into such territory. Instead, they will enjoy the response, the feeling of capability that a sports-car chassis like the 924's imparts.

Same for the 924's brakes, which with the help of the wide tires deliver powerful stopping forces to the ground and allow the driver good control of the car's direction in the process. Interestingly, the car's disc-drum combination also sails through our 6-stop fade test without any increase in pedal effort, an unusual feat for a car without discs all around. But 4-wheel discs are

available as an option should the buyer want the ultimate in brake specifications.

As mentioned earlier, early 924s suffered from a considerable amount of body noise on rough road surfaces; we recall being almost horrified by the resonant drumming inside on a typical European brick street, and just plain road noise on more normal surfaces seemed almost artificially amplified. Porsche's development efforts paid off well starting with the 1980 model; today one still gets the feeling of a rigid body with the wheels attached very firmly to it, but under most road conditions the 924 is now reasonably quiet.

A similar comment applies to engine noise getting inside the 924. The Audi-based four was always rough and noisy and from the outside it still answers that description. When you push it hard, it also gets loud inside. But in most normal driving it is now well damped, even almost refined in its sound as far as the car's occupants are concerned. With the latest combination of gearing, which has been the same since 1980, the 924 is a downright restful highway cruiser, its engine turning only 2400 rpm at 60 mph, and with its fairly generous fuel capacity of 16.4 gal. it's a pretty long-range highway cruiser. Our average in a good mixture of in-town, freeway and suburban driving was 22.5 mpg, and all-highway trips should give something in the high 20s.

Still, performance is not the 924's strong suit, at least not relative to its price. Power remains at 110 bhp—last year's *Guide* reported an increase to 115, but that turned out to be an error on the part of U.S. Porsche people—and this makes for lively if not brilliant acceleration. Last year's adoption of the 3-way catalytic converter and oxygen-sensor feedback, however, did do nice things for driveability; the 924's engine starts readily from cold and runs cleanly from then on.

From the standpoint of controls, the 924 communicates that directness, that intimacy with the machinery and the road that makes a sports car a sports car. The steering is precise and fairly quick; gearshifting is also precise, accomplished with a short lever with short throws and nice snick-snick action, comments that could not have been applied to early models. But there remains a high-inertia feel, traceable mainly to the fact that the gearbox synchronizers have to deal not only with gears but with the driveshaft as well.

The 924 Weissach's interior, with its durable-looking but comfy tweed seats and side panels, was one of the nicest of the many 924 cockpits we've seen yet. Its 2-tone tweed seats, shaped like those of other 924s (but "sports buckets" are available at extra cost), are firm and very supportive laterally although some drivers find them a bit too much of both. The 924's great eccentricity remains to bother most who drive it: the steering column wound up so low relative to the driver's seat that the designers mounted the steering wheel off-center on the column to allow the legs working room *most* of the time. One can correct this, after a fashion, by removing shims from the seat mountings. But the driver already sits pretty low in the 924, so not everybody is going to be willing to sit even lower.

Back to that price now. When compared with its two Japanese competitors, the Mazda RX-7 and Datsun 280ZX, the 924—Weissach or not—is working at a grave price disadvantage in the market place; its sales performance has reflected this for some time. The present high price, however, is based on a Deutsche Mark-to-dollar exchange rate that is no longer valid; in the past year, in fact, the Mark has lost nearly 40 percent relative to the dollar. Will Porsche lower the 924's price by 40 percent, then, to give the 924 a chance to sell in quantity?

Only time will tell. At this writing nobody is sure how long the dollar will remain so strong, so a wave of price reductions has yet to sweep across the imported-car field. But if the dollar holds, it could happen, bringing about one of the happiest developments for American car enthusiasts since we became inured to almost monthly price jumps. In the meantime, the 924 remains a very good, very roadable if not highly refined sports car—at a very high price.

PORSCHE 4-WHEEL-DRIVE 911 ROADSTER

New life for an old friend

BY PAUL FRERE

ORSCHE HAS BEEN working for some time on a 4-wheel-drive version of the 911, but its appearance at the Frankfurt Auto Show in the form of a 911 Roadster idea car came as a surprise. Actually, the new body style and the drive arrangement are two separate ideas. The non-Targa convertible will be added next year to the 911 line, but is unlikely to be offered to the public in combination with the 3.3-liter turbo engine and 4-wheel drive. The reasons are the same ones that have stopped Porsche from offering a targa version of the 911 Turbo, namely the inadequate torsional rigidity of the open car when combined with 300 bhp, and the company's belief that people who drive open cars don't need 300 bhp.

The idea of adding a drive to the front wheels of the 911 is

nearly two years old. It was probably triggered by Audi development chief Ferdinand Piëch, following his experience in developing the Quattro. Piëch is Dr Ferry Porsche's nephew, and the Porsche company is almost totally owned by Dr Porsche, his sister Louise Piëch (who looks after Porsche's interests in Austria) and the children. Of these, Ferry Piëch, 42, is one of the most brilliant engineers in the European automobile industry and as such gets a good audience with his uncle.

In a chat with Dr Porsche at the Frankfurt Show, he told me that as far back as 1979 he had asked Dr Ernst Fuhrmann, Managing Director of Porsche until the end of last year, to adapt 4-wheel drive to the 911. Fuhrmann, however, was anxious to kill the 911, which he had inherited from the previous management

The question is how solid the link should be between the front and rear wheels. Audi uses a differential to accommodate the speed difference between the front and rear wheels, but this causes unpleasant and power-consuming drag in sharp turns. However, the drive can be locked, which actually improves the high-speed handling and above all prevents any one set of wheels from locking prematurely during hard braking. Incidentally, the rear differential of the Audi can also be locked if maximum grip is required. Instead of the front/rear differential, Audi also tried freewheeling in its rally cars, but without much success. These solutions have also been tried by Porsche and the possibility of using a Ferguson mechanism providing an unequal torque split between the front and rear wheels is apparently also under investigation. [According to Peter Schutz a 40/60 front/rear torque split with a car of the 911's approximate 40/60 weight distribution is just about ideal.—*Ed.*] As far as could be ascertained, the roadster show car had freewheeling, but certainly no final decision has been taken yet as to which solution should finally be adopted.

Regardless of which answer Porsche uses, Dr Porsche says that the 4-wheel drive has solved all the problems related to the 911's handling, such as its strong understeer at moderate cornering speeds (to prevent excessive oversteer at the limit) and its sharp

team led by Dr Porsche himself, in favor of his babies, the 924 and 928, and soon shelved the project. Now with Peter Schutz in the Managing Director's seat, Dr Porsche again takes a much more active part in building company policy. One result is that the 4-wheel-drive project is being worked on by Helmuth Bott, who played a very important part in the development of the 911 from the beginning and was promoted to the board at Porsche some years ago to be in charge of technical development.

The basic design of the 911 makes it just as easy to convert to 4-wheel drive as it was in the case of the Audi, which has a similar mechanical layout turned back to front. At Porsche it is just a matter of extending the gearbox output shaft forward with a propeller shaft driving the front differential unit.

all-on or all-off reactions at high cornering speeds. However, it remains to be seen if adding the weight of the differential unit and the halfshafts to the front of a 911 would not have a similar or almost similar effect. I don't say this because I'm skeptical about 4-wheel drive, but in the early years of the 911, Porsche engineers found for themselves what sort of difference slight changes in weight distribution could make in a rear-engine car.

With more than 58 percent of its weight on the rear wheels (and more for the turbo or with air conditioning), the 911 is the last car that really needs 4-wheel drive, which also adds undesirable weight. But comparative tests on snowbound mountain roads have indicated that 4-wheel drive still makes a huge difference in such circumstances. There are many buyers, how-

CONTINUED ON PAGE 84

PORSCHE 944

More performance (and more Porsche) for the 924

BY PAUL FRERE

Although it's a stranger to our shores, the 944's chassis and body are essentially that of the 924 Carrera GT, a flared-fender, wide-wheel version of the standard sports/GT.

IT'S MORE OR less common knowledge (at least among the automotive cognoscenti) that the 924 is the car upon which Porsche will base much of its future line. Thus, in the seven years since its introduction, we've seen the 4-cylinder GT go through numerous evolutionary changes that are moving it upmarket to where it will ultimately replace the 911. That may take a while, says Porsche. In the meantime, the development of the 924 continues and the latest derivation, called the 944, has raised more than a few eyebrows.

Some of the new models unveiled at the 1981 Frankfurt Motor Show were entirely new while others were modifications of existing models. But almost all were engineered for better fuel economy without sacrificing performance.

The Porsche 944, however, is an exception as neither its performance nor its fuel economy is claimed to be better than the Porsche 924 Turbo's. The principal difference is one of image and development potential. It is well known that the 924's engine is basically an Audi design, and the engines are supplied by Audi though the turbocharged versions are actually assembled in Zuffenhausen. The 2.0-liter, 4-cylinder engine has not been used by Audi in a passenger car for two years, however, and some day Audi will stop making it. Arrangements have already been made with Porsche to ensure a supply for a 924 production run of about two years after official manufacturing has ceased, but the future doesn't warrant further development of the engine by Porsche.

Now Porsche has come up with its own engine, which has been dropped into the bodyshell of the 924 Carrera GT (with wide fenders and 7-in. wide rims) to make the 944. The 4-cylinder all-aluminum unit is virtually half of the 928's V-8, though the bore is slightly larger to increase the capacity to 2.5 liters. The Reynolds Aluminum bores are unlinered, and the distance between the centers is the same as in the V-8, so that the block and the head can be machined on the same line as the V-8s. The sintered metal connecting rods and most of the valve gear, including a single overhead camshaft and hydraulic tappets, are interchangeable with the V-8's.

Obviously, a 2.5-liter 4-cylinder unit peaking at 5800-plus rpm is a pretty rough piece of machinery. That's why Porsche didn't hesitate to repay the compliment the Japanese paid the European industry when they copied its motorbikes and cars: The manufacturer added two balance shafts at different heights on both sides of the cylinder block to balance the second-order shaking forces characteristic of inline-4 engines and at the same time to even out the engine torque reactions (the obvious reason for the different heights). The engine specialists say that the shafts don't absorb more than 3 bhp, leaving the engine to produce 163 bhp DIN at 5800 rpm (66 bhp/liter) and a modest maximum torque of 151 lb-ft as low down as 3000 rpm. The balance shafts are driven at twice crankshaft speed by means of a cog belt, while another cog belt drives the camshaft directly from the crankshaft. The balance shafts were used even in the special turbocharged version of this engine that made its competition debut at Le Mans in the modified 944 Walter Röhrl and Jürgen Barth piloted to 7th place overall. This racing engine also boasted a special twincam 16-valve cylinder head that, together with the turbocharger, pushed the power up to 410 bhp—a tribute to the toughness of the bottom end.

The slanted-roof combustion chambers have been developed one step further than in the latest V-8 engine and are notable for the squish effect obtained when the piston reaches top dead-center. They are so efficient that a compression ratio as high as 10.6:1 can be used with 98-octane fuel, the high compression ratio being assisted by the combined

The big show's on the inside, under the hood, where the familiar Audi-based 4-cylinder has been replaced by an all-aluminum sohc four that is half of a 928 V-8. With its slightly larger than stock 928 bore, this 2.5-liter powerplant produces 163 bhp.

Bosch L-Jetronic injection and full-electronic, flywheel-triggered ignition system with a programmed digital advance curve.

Except for slight differences in the gearing, the 944 is very similar to the 924 Carrera GT but it has a more comfortable suspension, although a sports suspension is optional. Another option is a 3-speed automatic to replace the standard 5-speed manual transmission.

Despite its sophisticated specifications the 944 engine is an immense unit for its 2.5-liter capacity. With 4.8 in. between bore centers, there is space for coolant all around the cylinders and ample room for further bore increases, so there would be no problem in raising the capacity to 3.0 liters.

I think a nice little 6-cylinder, either V or inline, would have been a much better proposition. Nor would I scorn a 924 using Audi's 5-cylinder engine. I will be very surprised if this is not the solution finally chosen to replace the current 4-cylinder in the basic 924.

75

PORSCHE 944

Worthy of the marque

PHOTOS BY JEFFREY R. ZWART

AT A GLANCE	Porsche 944	Datsun 280ZX Turbo	Alfa Romeo GTV 6/2.5
List price	$18,450	$17,299	$17,455
Curb weight, lb	2790	2995	2840
Engine	inline-4	inline-6	V-6
Transmission	5-sp M	3-sp A	5-sp M
0–60 mph, sec	8.3	7.4	9.1
Standing ¼ mi, sec	16.3	15.6	16.8
Speed at end of ¼ mi, mph	84.5	88.0	83.0
Stopping distance from 60 mph, ft	149	164	162
Interior noise at 50 mph, dBA	73	68	69
Lateral acceleration, g	0.818	0.754	0.778
Slalom speed, mph	60.9	58.6	62.0
Fuel economy, mpg	22.0	16.5[1]	19.0[1]
Issue		7-81	7-81

[1] Trip fuel economy

REMEMBER AESOP'S FABLE of the boy who cried "wolf" so often that when the real wolf appeared, no one heeded his cry? Well, there's a sort-of parallel in the automotive world and it goes like this. Once upon a time, a highly respected manufacturer of air-cooled, rear-engine sports cars introduced a new water-cooled, front-engine model known as the 924. "It is a Porsche," the builder told everyone, even though people knew that the car's engine was essentially an Audi and that much of the suspension and other mechanical and decorative tidbits were Volkswagen components. Yet, the people believed the manufacturer, at least until they had driven the car and found that although it looked and handled like a Porsche, it had an engine that was rough and noisy and a ride that was jouncy and boomy. So they told the constructor who then made some improvements (but no real changes) and continued to call the 924 a Porsche. The car-buying public stopped listening and when other manufacturers took up the cry (in Japanese, the equivalent for the word Porsche sounds like Datsun and Mazda), people turned their ears to the Orient. Soon, these enthusiasts were helping keep the wolves away from Nissan's and Toyo Kogyo's doors by buying Z-cars and RX-7s rather than 924s.

Late in 1981 the German car company again cried "Porsche" when it introduced the latest derivation of the 924 to the European motoring public. But lest people not believe the alarm, the factory quickly pointed out that the new model was actually a 944 and that it had a new engine as well as a restyled body and revised suspension. "Porsche," cried the manufacturer. "Porsche," echoed the public, which soon discovered that here indeed was a car worthy of its name and reputation. And that's no fable!

We'll forgo any further allusions to Aesop and say simply that the 944 is one of the most exciting cars to come out of Zuffenhausen (by way of Neckarsulm) in a long time. It is spirited, yet smooth. It handles superbly, yet its ride is not harsh. It has swoopy looks, period. "Super! Looks great with those wide fender wells," exclaimed one staffer. Whether or not you agree depends on your personal tastes in styling. If you've a penchant for understated looks then it may take some convincing to get you to like the 944. Driving one is a sure-fire way to do so.

Anyone who remembers the old Audi-based 924 4-banger will really appreciate the new sohc inline-4. It's mostly aluminum and if it looks familiar it's because it is one half of the V-8 used in the 928. You see, in 1976 when Porsche was searching for a suitable replacement for the 924 powerplant, the factory considered the obvious alternatives. Inline-5 and 6-cylinder engines were too long and too heavy and existing V-6s such as the Renault/Volvo/Peugeot engine were "not encouraging." Porsche felt a 4-cylinder was still the best choice because of fuel efficiency and size—the engine had to be compact enough to be installed from below on the Audi assembly line. The existing four, given a shot in the arm through turbocharging, was deemed too heavy, not overly suitable mechanically and too rough. No, what was needed was a new, lightweight, smooth-running 4-cylinder.

Looking at their inventory, Porsche engineers realized they had the makings of the right engine in the aluminum 928 V-8. Why not cut it in half longitudinally and make it a four? Fine, that would certainly take care of excess weight. But what about smoothness? Wouldn't the new inline 4-cylinder have the same secondary imbalances and thus rough running as the 924 powerplant? Not if balance shafts were used.

Although the balance-shaft concept, the Lanchester principle, dates back to 1911, its most widespread application came in the Sixties with Mitsubishi engineers using two counter-rotating shafts running at twice engine speed. What's more, the Mitsubishi design contained a patentable feature: using only two bearings per shaft to minimize frictional losses. Porsche tried to circumvent the patent by fitting three bearings, but finally decided that it was less expensive to pay royalties than to spend money on further development.

In choosing the half V-8 design, Porsche has kept costs down because the two engines share some of the same basic components. For example, connecting rods are machined from the same blanks, although the finished products are not interchangeable. This universality is possible because the two engines have the same 78.9-mm stroke. But each powerplant has a different bore (the 944's is 5.0 mm larger) so the pistons are not identical.

Both engines use fuel injection, but while the 928's is the relatively simple Bosch L-Jetronic, the 944's injector uses a state-of-the-art Digital Motor Electronic system that offers (among other things) precise control of fuel flow and ignition. Suffice to say, it offers the best of all worlds—good performance and driveability as well as optimum fuel efficiency and low emissions. More about performance later, after we examine the 944's styling and suspension.

Flared fenders, a sweeping front spoiler with air scoops and built-in driving lights, plus 911-style 5-bolt wheels are a few of the styling traits that set the 944 apart from even a customized 924. Yes, the new car's fenders are similar to some of the aftermarket flares one finds on 924 cafe racers, competition cars and the 924 Turbo Carrera GT. But all of these are fiberglass add-ons, whereas the 944 fenders are steel (zinc-coated and guaranteed against rust for seven years) and form an integral part of the body. The front spoiler is polyurethane and conforms to the contours of the fenders. The rear spoiler attaches to the all-glass rear hatch and is identical to the 924 Turbo's. Although this outrageous bodywork imparts a racy look to the 944, it has a more useful function: streamlining. Porsche claims the car has a drag coefficient of 0.35.

Now then, about those wheels. They may look like the 911's but they are not the same. The offset is different although the bolt pattern is not. Anyway, standard 944 issue is the 15 x 7-in. cast alloy cookie cutter design shod with Pirelli CN36 215/60VR-15 radials. The 16 x 7-in. forged alloy wheel fitted with 205/55VR-16 Pirelli P7s is optional.

Beneath the bulging bodywork is the familiar 924 suspension: MacPherson struts with coil springs up front; semi-trailing arms and torsion bars at the rear. Porsche says that there are no drastic changes in the basic setup. Instead such things as spring rates, anti-roll bar diameters and shock absorber damping have been honed to provide optimal handling and ride. Nor are there any changes in the 4-wheel disc brakes that are the same as those of the 924 Turbo, and (here we go again) similar to the 911's discs.

And we would be derelict if we didn't mention the lengths to which Porsche went to isolate engine and driveline noise and vibrations in the 944. The factory designed an unusual aluminum crossmember to support the engine and suspended the entire package in the unit body by using two specially designed hydraulic engine mounts. Two conventional rubber mounts were used at the rear of the car where the transaxle is located, providing a total of four suspension points for the engine/driveline module. And to dampen vibrations at the steering wheel, Porsche engineers used rubber bushings to secure the steering rack to the car's front crossmember.

Climbing out from under the car and sliding into the driver's seat, we find that not much has changed. The design of the seats, dash and console is familiar 924 fare, which is comfortable, efficient and pleasant. If you're waiting for us to say, "and Teutonically austere," don't. The latest interior is inviting in a sporting sort of way. It's all black with a leather-like material covering the door panels, dash and center console. The carpeting that extends into the rear trunk is also black and of deep pile. Our test car had the optional leather sport seats (comfy with excellent side bolstering) and they were black too. However, you can order tweed seats instead. Although the instrument design and layout are the same as before, the new car's dials and gauges have yellow numerals placed over a black background. Some people may consider yellow too garish, but the idea is maximum visibility and, by God, these are very readable. Especially the tachometer, which starts its swing at 3 o'clock and rotates clockwise to its 6400-rpm redline (yellowline, actually) at about 11.

The Porsche name is embossed on both windowsills and there's a Porsche crest on the glovebox latch. Turn it and expose the keyhole located beneath it. Touch the two rocker switches on the driver's door to raise and lower the windows. Tap the wobble button on the left armrest and adjust either electrically actuated and heated side mirror. These niceties as well as air conditioning, a removable sunroof, tinted glass, a 3-spoke leather-covered steering wheel and a leather-capped shift lever are standard on the 944, a car that's chock-full of features and other surprises. Unfortunately, there's one major disappointment, the position of the steering wheel. In case you've forgotten, the wheel is low so one encounters hand-to-thigh interference when turning the wheel more than 90 degrees. We noted that Porsche tried a quick fix, leaving out the spacers under the seats. But this just complicates matters because it reduces the seating height and makes it difficult to see over the rear spoiler. We're hoping that ultimately Porsche will attend to this problem. In the meantime, consider it a minor annoyance because the rest of the car is so nice we can live with a less than ideal steering wheel position.

If, after checking out the details of the interior, you fasten your seatbelt and turn on the ignition, you'll find that the engine fires up immediately and runs smoothly, even when cold. And Lordy, does it rev—right up to the redline in every gear except 5th. There are no stumbles, flat spots or resonance points. Furthermore, there's low- and mid-range flexibility that allows you to drop the revs to as low as 1000 rpm in top gear and the engine pulls without protest.

With 143 bhp on tap, the 944 powerplant gets you from 0 to 60 mph in 8.3 seconds and to the quarter-mile in 16.3 sec. That's quicker than the 924 Turbo, which has the same horsepower but takes 9.3 and 17.0 sec to reach the same speed/distance. The 944 is tops in top gear too. Given its head, it will reach 132 mph or about 3 mph more than the 924 Turbo. Those are mighty impressive numbers but they are only part of an engine package that guarantees the sort of smoothness one expects only from an inline-6.

You could easily while away your hours enjoying the new engine's performance and smoothness. But you'd be missing the other half of the 944's exciting nature, its handling. The 944 beat two of its evident competitors, the Datsun 280ZX and Alfa Romeo GTV 6/2.5, very handily around our skidpad, though the Alfa was a bit more nimble through our slalom. In sibling competition, it more than holds its own with the 924 and even the 911SC. However, in our tests, it was not quite as good as the 924S Turbo because unlike that car, our 944 was equipped with standard-issue 60-series tires and not the 55-series Pirelli P7s. This made a difference of about a tenth of a g and 1 mph on the skidpad and in the slalom, respectively. But given a similar, wide-footprint, sticky tire pairing, we'd bet the 944 would surpass the Turbo in both tests.

This may be comforting news to Solo II slalom racers, but what does it mean to the everyday driver? That the 944's combination of power, mid-range torque, flexible gearing, superb handling and excellent braking result in an extremely well balanced machine. The 944 has a neutral feel and a tendency toward power understeer. Throttle liftoff brings the tail out, but ever so slightly and controllably. Knowing this, an experienced driver can play throttle, brakes and steering to wring the maximum out of the car.

If all of these accolades seem effusive, it's because the 944 is the sort of car that makes driving fun. A top-notch GT, it's what the 924 should have been and wasn't. It seems that Porsche thinks so too because in what looks like an attempt to make us forget the past, the company is dropping the 924 and the 924 Turbo from the North American lineup. That's right, the 944 will become Porsche's lowest-priced model and one of the best sports car bargains. According to Porsche-Audi, the introductory price will be $18,450 base. That buys you a lot of engineering, performance, handling and looks, not to mention a heck of a lot of Porsche.

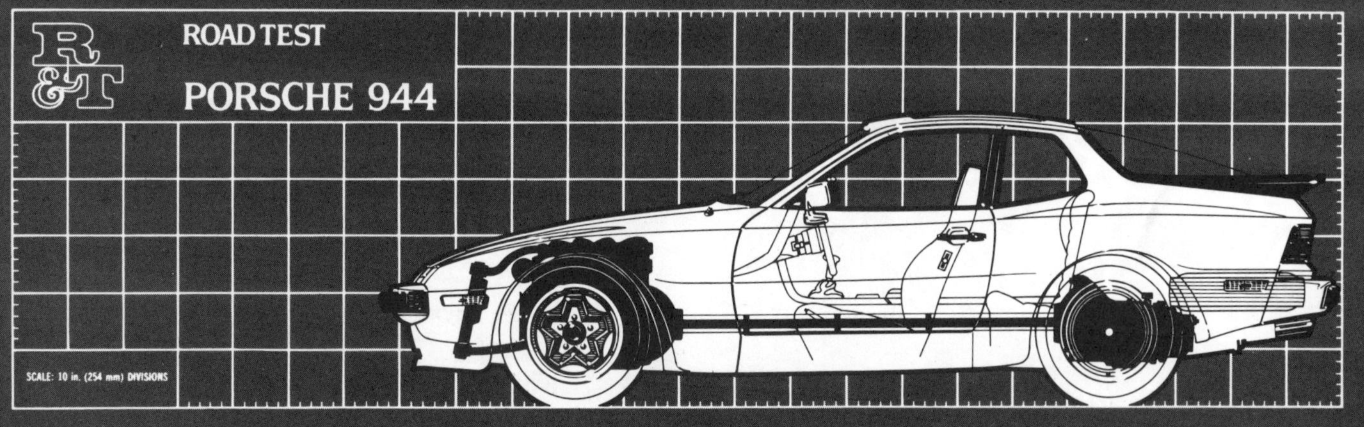

PRICE

List price, all POE $18,450
Price as tested ....est $20,000
  Price as tested includes std equip (air cond, sunroof, alloy wheels, elect. window lifts, elect. adjustable and heated side mirrors, elect. antenna, foglights), AM/FM stereo/cassette and sport seats (est $1550)

IMPORTER

Porsche-Audi Div, Volkswagen of America, Inc, 818 Sylvan Ave, Englewood Cliffs, N.J. 07632

GENERAL

Curb weight, lb/kg 2790 1267
Test weight 2945 1337
Weight dist (with driver), f/r, % 49/51
Wheelbase, in./mm 94.5 2400
Track, front/rear ....58.2/57.1 ...1478/1450
Length 170.0 4318
Width 68.3 1735
Height 50.2 1275
Ground clearance 4.9 124
Overhang, f/r ....37.4/41.1 .... 874/1044
Trunk space, cu ft/liters .... 10.4+7.9 ..... 295+224
Fuel capacity, U.S. gal./liters 16.4 62

INSTRUMENTATION

Instruments: 85-mph speedo, 6400-rpm tach, 99,999 odo, 999.9 trip odo, oil press., coolant temp, fuel level, clock
Warning lights: oil press., brake system, handbrake, alternator, low fuel, oxygen sensor, rear-window heat, seatbelts, hazard, high beam, directionals

ENGINE

Type ....sohc inline-4
Bore x stroke, in./mm .....3.94 x 3.11.. 100.0 x 78.9
Displacement, cu in./cc 151 2479
Compression ratio.........................9.5:1
Bhp @ rpm, SAE net/kW............. 143/107 @ 5500
  Equivalent mph/km/h 134/216
Torque @ rpm, lb-ft/Nm............. 137/186 @ 3000
  Equivalent mph/km/h 75/121
Fuel injection........................Bosch L-Jetronic
Fuel requirement ....unleaded, 91-oct
Exhaust-emission control equipment: 3-way catalyst, oxygen sensor

DRIVETRAIN

Transmission ....5-sp manual
Gear ratios: 5th (0.73) 2.84:1
  4th (1.07) 4.16:1
  3rd (1.46) 5.68:1
  2nd (2.13) 8.29:1
  1st (3.60) 14.00:1
Final drive ratio 3.89:1

ACCOMMODATION

Seating capacity, persons ....2+2
Head room, f/r, in./mm.....37.0/32.5 ......940/825
Seat width, f/r....2 x 19.5/2 x 15.0..2 x 495/2 x 381
Seatback adjustment, deg 50

MAINTENANCE

Service intervals, mi:
  Oil/filter change 7500/7500
  Chassis lube ....none
  Tuneup...................................15,000
Warranty, mo/mi.........................12/unlimited

CHASSIS & BODY

Layout ....front engine/rear drive
Body/frame ....unit steel
Brake system ...11.1-in. (282-mm) vented discs front, 11.4-in. (289-mm) vented discs rear; vacuum asst
  Swept area, sq in./sq cm........ 446......... 2878
Wheels ....cast alloy, 15 x 7J
Tires ....Pirelli CN36, 215/60VR-15
Steering type ....rack & pinion
  Overall ratio 22.4:1
  Turns, lock-to-lock 4.0
  Turning circle, ft/m .... 31.2............ 9.5
Front suspension: MacPherson struts, lower A-arms, coil springs, tube shocks, anti-roll bar
Rear suspension: semi-trailing arms, torsion bars, tube shocks

CALCULATED DATA

Lb/bhp (test weight) 20.6
Mph/1000 rpm (5th gear) 25.0
Engine revs/mi (60 mph) 2400
Piston travel, ft/mi 1245
R&T steering index 1.25
Brake swept area, sq in./ton 303

RELIABILITY

Owners of earlier-model Porsches reported 11 problem areas and 4 disabling reliability areas compared to overall Owner Survey averages of 12/6. So we expect the overall reliability of the 944 to be average.

ROAD TEST RESULTS

ACCELERATION

Time to distance, sec:
  0–100 ft 3.2
  0–500 ft 8.9
  0–1320 ft (¼ mi) 16.3
Speed at end of ¼ mi, mph 84.5
Time to speed, sec:
  0–30 mph 2.5
  0–60 mph 8.3
  0–80 mph 14.3
  0–100 mph 23.9

SPEEDS IN GEARS

5th gear (5400 rpm) 132
4th (6400) 109
3th (6400) 80
2nd (6400) 55
1st (6400) 33

FUEL ECONOMY

Normal driving, mpg 22.0
Cruising range, mi (1-gal. res) 339

HANDLING

Lateral accel, 100-ft radius, g.....0.818
Speed thru 700-ft slalom, mph 60.9

BRAKES

Minimum stopping distances, ft:
  From 60 mph 149
  From 80 mph 256
Control in panic stop ....excellent
Pedal effort for 0.5g stop, lb 19
Fade: percent increase in pedal effort to maintain 0.5g deceleration in 6 stops from 60 mph 42
Parking: hold 30% grade? ....yes
Overall brake rating..........excellent

INTERIOR NOISE

Idle in neutral, dBA 58
Maximum, 1st gear 77
Constant 30 mph....................68
  50 mph.........................73
  70 mph.........................80

SPEEDOMETER ERROR

60 mph indicated is actually 60.0

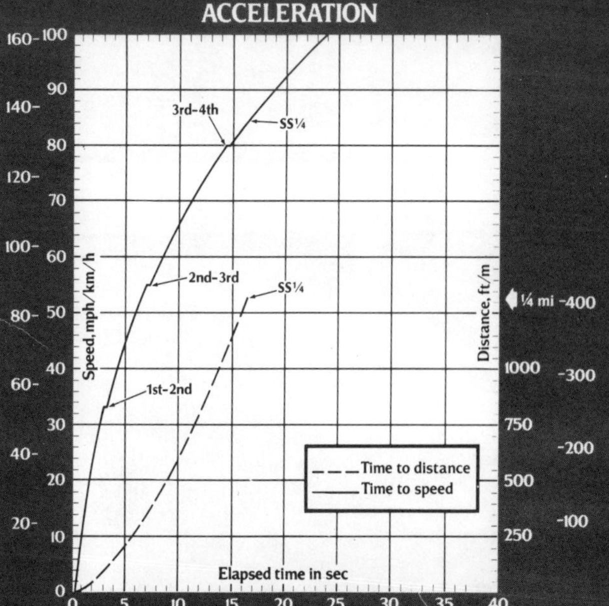

PORSCHE 914 & 914/6 1970-1976

Cheap Porsche or expensive VW, the driving fun is abundant

BY THOS L. BRYANT

USED CAR R&T CLASSIC

PHOTOS BY GORDON CHITTENDEN

BRIEF SPECIFICATIONS

	1970 914	1970 914/6	1976 914 2.0
Curb weight, lb	2085	2195	2250
Wheelbase, in.	96.4	96.4	96.4
Track, f/r	52.6/54.1	53.6/54.4	52.8/54.4
Length	156.9	156.9	164.4
Width	65.0	65.0	65.0
Height	48.0	48.0	48.0
Engine type	ohv flat-4	ohc flat-6	ohv flat-4
Bore x stroke, in.	3.24 x 2.59	3.15 x 2.60	3.70 x 2.80
Displacement, cc	1679	1991	1971
Horsepower @ rpm	85 @ 4900	125 @ 5800	84 @ 4900
Torque @ rpm	103 @ 2800	131 @ 4200	97 @ 4000

PERFORMANCE DATA
From Contemporary Tests

	1970 914 1.7	1970 914/6	1976 914 2.0
0-60 mph, sec.	13.9	8.7	12.7
Standing ¼ mi, sec.	19.2	16.3	19.2
Avg fuel consumption, mpg	25.5	21.3	26.5
Road test date	4-70	7-70	1976 S>

TYPICAL ASKING PRICES*

1970–1973 914 1.7	$3000–$5000
1970–1971 914/6	$8000–$10,5000
1973–1974 914 2.0	$4000–$6000
1974–1975 914 1.8	$4000–$6500
1975–1976 914 2.0	$5000–$7500

*Prices are estimates based on cars that are in reasonably good condition but not restored to like-new. Cars in excellent condition and/or fitted with optional equipment will, of course, command higher prices, while poorly maintained cars may be cheaper than the range given here.

POTENTIAL PROBLEM AREAS
From November 1974 R&T Owner Survey

Problem areas reported by more than 10 percent of owners: instruments, body parts, upholstery, engine mechanical, fuel injection, clutch cable, starter, gearbox.

Problem areas reported by 5–10 percent of owners: fuel lines, windows, distributor, fuel pump, heater-ventilation.

SOMEHOW IT SEEMED more important in 1970 to distinguish between the Porsche 914s and the *real* Porsches, i.e., the 911 and 356 models. The 914 was treated rather rudely by most Porsche enthusiasts in the U.S., and for years there were many Porsche clubs that refused to open their doors to 914s. In Europe, it was sold under the combined name Volkswagen-Porsche 914, but the U.S. distributor wanted no part of that moniker, although it did signify the joint marketing agreement signed by the two German automakers. Now, 12 years later, the controversy over whether or not the 914 is truly a Porsche has lost much of its steam, particularly in light of the unveiling of such models as the 924 and 928. Nonetheless, there are those who continue to disparage the 914. The prospective Used Car Classic buyer who is status conscious may want to consider selecting another sports car. But if measuring yourself against someone else's yardstick isn't as important as driving fun, read on.

The first public viewing of the mid-engine 914 and 914/6 came at the Frankfurt Auto Show in the autumn of 1969, and the considerable interest in a new model from Porsche, the first of the modern sports cars with such a layout, was tempered by the car's styling. Criticisms focused on the car's slab sides and pinched-fender front end. On the other hand, the rear styling was rather more acceptable, and the targa-style removable roof was applauded as an efficient blend of open and closed motoring. The interior appointments were quite plain compared to the 911 models, and the less-than-luxurious seats came in for considerable criticism, although they actually did provide reasonable lateral support and fair comfort. As with *real* Porsches, there was abundant head, leg and elbow room in the 914.

A differentiation was made by most Porsche enthusiasts between the 914 and the 914/6—the latter being more acceptable as nearly the real thing. Whereas the 914 used primarily VW running gear and powertrain, the 914/6 (which came to the U.S. some three months after the 914) was built with more pieces out of the Porsche parts bin, which is only fitting as it was assembled at Zuffenhausen, the Porsche plant outside Stuttgart, while the 914 was put into finished form at the VW plant in Wolfsburg.

The 914 was fitted with VW's flat-4 from the 411LE sedan and featured Bosch electronic fuel injection. This 1.7-liter (1679-cc) powerplant boasted a fairly modest 85 bhp at 4900 rpm and 103 lb-ft torque at 2800, resulting in performance that was characterized as somewhat disappointing. In our initial road test (April 1970), we measured a 0-60 mph time of 13.9 seconds and a quarter-mile capability of 19.2 sec at 70.0 mph. The transaxle was a 5-speed manual which some observers felt was a bit pretentious for the engine's performance, and both 4th and 5th gears were overdrive ratios (0.927 and 0.708:1, respectively).

The 914/6 was considerably more powerful than its sibling and featured larger brakes, wider wheels and tires and more instrumentation. The engine was the 911T's 2.0-liter (1991-cc) opposed-6, rated at 125 bhp at 5800 rpm and 131 lb-ft torque at 4200. Two Weber carburetors were used and the performance was such that in our July 1970 road test we stated that it "is such a beautiful, strong engine that it transforms the 914 back into a real Porsche—say what you will about the chassis, the 4-cylinder version still comes off as a VW." In hard numbers, the 914/6 demonstrated that 2.0 liters, properly done, can be enough as it

sped from 0-60 mph in 8.7 sec and romped through the quarter-mile lights in 16.3 sec at 83.0 mph—just a few ticks slower than the 911T's time of 16.0 sec.

The testers of the day were even less enchanted with the wisdom of a 5-speed in the 914/6, pointing out that the 911T was able to manage quite well with a 4-speed gearbox. They did acknowledge that the shift linkage in the 914/6 was "much better than the vague operation we experienced with the 4-cylinder. There's a different engine layout to pass the linkage around as well as the fact that Porsche assembles the 914/6 (the whole car) while VW builds the four." An automatic gearbox, the Porsche Sportomatic, was also offered but very few buyers elected that option.

In discussing the handling attributes of the 914, our initial skid-pad number was a disappointing 0.723g. The report added, however, that "its transient characteristics—the way it responds when you first steer it into a turn or change the throttle opening in the middle of a turn—are excellent. Initial response to a steering input is utterly without delay, a feature we expect with a central engine; body roll is so slight as to go unnoticed. And what happens when the driver lifts his foot from the throttle in a hard corner—this is the trickiest thing about rear-heavy cars—is simply a mild tuck-in of the front or, at the extreme, a smooth breakaway of the rear. Even if the driver does something stupid (like stabbing the brakes) after finding himself in a corner going too fast, the 914 does nothing violent and control can be recovered easily."

Very few changes were made to the 914 and 914/6 for 1971 (the 4-cylinder car's interior and trim were made slightly fancier). That year marked the end of the line for the 6-cylinder model after a production run of some 3107 cars. Porsche cited price resistance (it cost almost as much as a base 911), and complications from increasingly stringent emissions regulations as reasons for putting the 914/6 to rest.

For 1972, the 914/4 sailed on alone with little alteration: The ventilation system was improved by adding fresh air ducts at each end of the dashboard, the quality of the carpeting was upgraded, and the passenger's seat was finally made fully adjustable, something that should have been done from the beginning.

By 1973, U.S. emissions standards were beginning to seriously affect the 1.7-liter engine's performance: The 49-state version was now showing 76 bhp with the original compression ratio of 8.2:1, but in the crucial California market, even tougher standards meant special tuning that resulted in an output of 69 bhp and a compression ratio of 7.3:1—scarcely in keeping with the car's image and price (the base price had risen from $3595 in 1970 to $4749 for 1973). To counter this situation, Porsche (or VW) introduced the 2.0-liter 4-cylinder version that boasted 91 bhp at 4900 rpm. The 914 2.0 package also included more attention to interior appearance, a center console that housed a voltmeter, clock and oil temperature gauge, forged alloy wheels and front and rear anti-roll bars.

Outside, the 1973 models were fitted with rubber bumper guards that marked the first major exterior change since introduction. The next year, federally required safety bumpers gave the 914 its only significant change in appearance. The anemic 1.7 engine gave way to the 1.8-liter (1795-cc) that was a 50-state ⟫⟫→

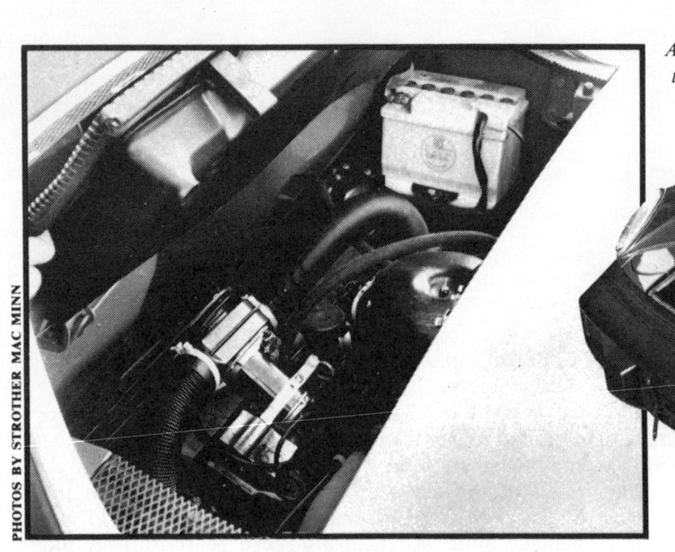

Access to the engine is through the small hatch behind the rollbar.

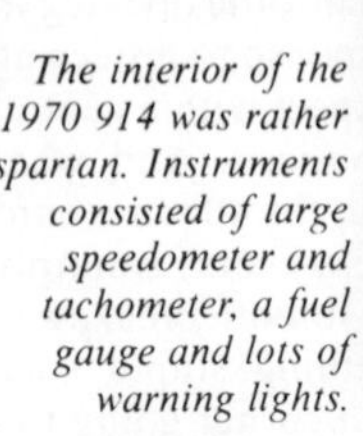

Battery perch (above) was not a good idea and leaking acid could result in ruptured fuel lines, leading to engine compartment fires.

The interior of the 1970 914 was rather spartan. Instruments consisted of large speedometer and tachometer, a fuel gauge and lots of warning lights.

Storage compartments front and rear gave the 914 more luggage space than most mid-engine designs.

engine and produced 76 bhp. The 2.0-liter was unchanged except for emission controls that reduced its output to 88 bhp.

That year (1974) also marked the "boutique" approach to marketing cars, and 914 buyers could select from a wide variety of new paint colors and special graphics. There was another change in bumper design for 1975 (and many observers found this treatment the most visually pleasing of all), but otherwise the 914 continued unaltered. A front spoiler could be ordered as an option and fitted below the front bumper, and the alloy wheels were now optional on both the 1.8 and the 2.0.

The 914 line came to an end with the 1976 model year—some 2686 versions of the 2.0 were built that year, but in May production at the Karmann coachworks was halted. R&T's Motor Sports Editor, Joe Rusz, noted in his book *Porsche Sport*

1976/77 that the "914's life was short but prolific. It was the most popular Porsche ever sold—83,841 units (U.S.) in seven years. And it was controversial—a Porsche to some, a Volkswagen to others. Ironically its successor, the front-engine 924, seemed destined for the same life of controversy."

Buying a Used 914

ENGINE COMPARTMENT fires are the first thing many people think about in considering Porsche 914s. There was a serious problem with rupturing fuel lines (often caused by the battery tipping over or coming unfastened and leaking onto the fuel lines that were just below the battery's platform in the engine bay). A car that shows evidence of such a fire may be a reasonable purchase, but you must ascertain if the fire was sufficiently serious to have damaged or weakened the car's frame.

With the help of more than 300 R&T readers who completed our 914 Used Car Classic Questionnaire, we've gained a comprehensive picture of the major areas of concern in seeking the 914 of your dreams. As with most sports cars there are those examples that have been driven into the ground and those that have been accorded kid-glove treatment. For some reason a lot of 914s were not well cared for, but this means that there is a fairly good supply of spare parts available in wrecking yards around the country where 914s have been laid to rest.

Rust is one of the major problems with any older sports car and the 914 is no exception. The cowling between the top of the front fenders and the windshield is prone to rust, as are the rocker panels and the bottoms of the doors and the sills. The rear suspension pickup points and the passenger's floor, along with the area where the roof pillar joins the body, are other locations where rust takes hold. A Richmond, Virginia owner wrote that he had rust spots around the windshield, fenders, doors, sills and headlights despite garaging his car and rarely driving in snow.

The pop-up headlights can be a problem in 914s, and they are not inexpensive to replace. If the car has suffered front-end damage the price of repairs may be prohibitive if you're not a competent do-it-yourselfer. The valance panels under the front and rear ends are susceptible to damage from parking lot bumpers, snowbanks, and whatever debris one may stumble upon, and it's relatively common to see 914s being driven without them, unsightly though that may be.

Inside there aren't too many serious problems with the 914, thanks in large part to the original simplicity of the design. The carpeting, particularly in the earliest models, is prone to wear. Also, it's not uncommon to find the window cranks broken, but these are VW parts and are readily available, as are the switches, knobs and levers.

In looking over the mechanical problems reported by our questionnaire respondents, we find that the fuel line difficulties previously mentioned are among the most common. Many owners have resorted to aircraft-quality fuel lines to prevent rupturing. As with other air-cooled Porsches, there is also the frequent problem of exhaust system heat exchangers wearing out, and replacement isn't cheap. Other relatively common problems are broken clutch cables, door handles and instrument failures.

As is true of other Porsches and most German cars, the parts and repairs for the 914 and 914/6 are rather expensive. Because of its exclusivity and performance advantage, the 914/6 is considered a more desirable car. Those examples of this model that are still around are usually kept in good condition and will be offered for sale at fairly high prices. But, as one Pittsburgh, Pennsylvania owner noted, "What a fun car—you have to drive it to believe it!"

Driving Impressions

JEFF ZWART, photographer extraordinaire whose work appears in our pages and on our covers from time to time, is the second owner of a 914/6 that was picked up at the factory by the original buyer. The flared fenders are steel and were a factory option, but not a widely advertised one and thus rather rare. Jeff has owned the car for 12 years and says he prefers its handling to that of the 911 he formerly owned. His car is also fitted with the aftermarket Richie Ginther suspension kit that included Teflon bushings, special coil springs in the rear and a 22-mm front torsion bar.

With more than 130,000 miles on it and the engine never apart, Jeff characterizes his car's performance as still brisk but showing signs of tiredness. In my brief time at the wheel, I found it quick and exciting, with plenty of mid-range torque and lots of top-end speed. The 5-speed gearbox is beginning to exhibit synchro wear, but it gets the job done, albeit with a feeling of a long distance between the shift lever and the gearbox itself.

The handling characteristics of the 914 have always been entertaining, and it's a car that can be driven quickly on first acquaintance. It has the proverbial "on-rails" feel and a marvelous responsiveness that encourages you to press on, faster and faster in the corners. Jeff had added Scheel seats to his car, and while they do offer greater support they also take away a bit of the roominess.

The controversy about the 914 will probably continue as long as the car itself is on the road—is it or is it not a real Porsche? I suppose each person has to answer that question for himself, but the fact is the 914 and 914/6 are great fun to drive. And isn't that the basis for selecting a sports car?

Rear transaxle, looking toward front counterpart.

Front bell housing, looking toward rear.

New notch for halfshaft lies ahead of one for steering linkage.

CONTINUED FROM PAGE 74

ever, who will not wish to put up with the additional weight (about 130–140 lb) and costs, and when the 4-wheel-drive 911s are put on the market, they will definitely be just alternatives to the normal models.

As for the roadster body, there is very little reworking of the Targa needed to make it a full convertible. The rollbar doesn't add appreciably to the strength of the body and so only a few minor added gussets are needed to make the conversion. Although the 4wd car has the flared fenders of the Turbo model, the production roadster will have those of the standard 911 series.

For 911 fans the Frankfurt show car was very important because it signaled just how long a life the 911 still has, one that wouldn't have been as obvious at the last Frankfurt show two years ago. Now Porsche is willing to take the time and effort to produce a roadster 911 and redo the model's drivetrain for 4-wheel drive. Recently the factory also prepared a brand-new 911 rally car for World Rally Champion Walter Röhrl to drive in the San Remo Rally.

Who would buy the 4-wheel-drive 911s? Private rally teams, because the continent is rally crazy now and the new car would offer Porsche a chance to get a potential winner into competition while possibly avoiding the enormous costs of running its own team. Then there are the skiers, this sport being very popular with Porsche owners who could justify the added weight and costs. And a 4-wheel-drive 911 is a marvelous way to take on snowy Alpine passes.

Profile:
PETER W. SCHUTZ

*American
at the Porsche helm*

BY RON WAKEFIELD

IT WAS AT the very least a mild shock for Porsche employees and German auto enthusiasts when a German-American named Peter W. Schutz was named Chairman of the Board of Porsche. For Professor Ernst Fuhrmann, the former chairman who had to retire early to make way for the new management, it must have been more than a mild shock. But things were not looking very good at Porsche. The 928 was not being accepted as the "ultimate" Porsche. Sales of the 924 and its variants were going steadily downhill. Only the faithful old 911 was holding its ground.

What to do? Through a management recruiter, Dr Ferry Porsche contacted Schutz, who had returned to his native Germany in 1978 to become a director of a truck manufacturer in Cologne. Their first talks were exploratory; there were several long sessions before a job offer was tendered. "Dr Porsche and the Piëch* family put a lot of emphasis on maintaining the high spirit of this company," Schutz recalls, "on making it a place where people can develop personally rather than just doing a job. I think it was the drive to rebuild some of that feeling, which was beginning to slip away. They were also looking for an engineer, as well as someone with a feel for the market place and the customers, experience in international business and particularly knowledge of the American market."

Peter Schutz meets all those criteria. Born in Berlin in 1930, he left Germany in 1938 with his Jewish family. After waiting two years in Cuba for a U.S. visa, they moved to Chicago, where Peter's father fulfilled the requirements for returning to his pediatric practice and the two Schutz boys aimed toward engineering careers.

Whereas his older brother wound up with a large paper-products company, Peter was clearly cut out for things automotive. He bought his first car, a 1937 Dodge, before getting his driver's license and overhauled "that rust bucket" with the help of several friends at the service station where he worked. The next car was a 1940 LaSalle coupe—with its 354-cu-in. V-8, "a Ford-eater." A 1953 DeSoto with a hemihead V-8 followed.

The only non-American car Schutz owned before coming to Porsche was a 1952 MG TD that he rebuilt completely—"and constantly. It was a mechanical nightmare," he recalls, a factor that probably helped lead to his next enthusiasm: flying. In any case, the MG was replaced by a Cessna 170 after Schutz got his pilot's license in 1962. A succession of planes followed; from 1962 until 1966, Schutz was an avocational partner in a Peoria flying school and cars took a back seat to the winged machines. His next car "of consequence" was a 1970 Chevrolet Malibu convertible, which he still owns, and the last automobile before Schutz entered the world of German company cars was a 1976 Corvette that he only recently sold.

As a teenager, Schutz had earned his first dollar working at that service station in Chicago during off-hours from high school. His next educational step was a Bachelor of Science degree in mechanical engineering from the Illinois Institute of Technology in 1952, after which he went to work for the Caterpillar Tractor Co as a test engineer, was drafted into the army for a 2-year stint, and returned to Caterpillar. In 1966 he joined the Cummins Engine Co, where he worked his way up to be a vice president.

The return to Germany came about when a management consultant recruited him for the truck-manufacturing job. Was it like returning home?

"No," answers Schutz, who left Germany at such an early age that he has had to relearn German, "there's no place to live like the U.S. I'm in Germany because this is the best job in the world for me." Peter and his second wife Sheila live in the Stuttgart suburb of Maichingen, where they not only feel at home but enjoy a steady stream of visitors from America. Among them are Shutz's children by his first marriage: identical twins Michael and Mitchel, 23 and both pursuing technical careers, and daughter Lori, an entomologist, who is 25.

A family man, then, and American through and through. "He wants us all to use the German *du* form with him," says Porsche press chief Manfred Jantke. That's virtually unheard-of in German management, where it's not unusual even for colleagues who have worked together for decades to stick with the *Sie*, or formal, form of "you" in dealing with each other. Jantke adds that Schutz himself goes ahead with the familiar form, but that the Germans around him can't quite bring themselves to reciprocate.

Other elements of his American management style are equally eyebrow-raising for Schutz's German co-workers. Even the term "co-worker" is significant, as he takes his top-of-the-heap position far less seriously than is customary in Germany. Instead of rank, he emphasizes openness, accessibility and plenty of information for Porsche employees.

Schutz, who before joining the company knew relatively little about Porsche, has learned very fast and is clearly enjoying himself. His first year has been devoted mainly to short-range decisions and correcting what appears to have been a certain lack of direction at Porsche. One of his first and easiest major decisions was to put the 911 back on the front burner. A first result was seen at the Geneva Salon: the production 911SC Cabriolet. Four-wheel drive is coming, probably in 1984, for the 911. And the 911 body—whose aerodynamics are no longer as good as those of the best contemporary sedans—may eventually get a thorough rework.

With the front-engine models, Schutz has concentrated on "cleaning up the evolution," as he puts it. A 944 Turbo is in the works, and with the 944 already replacing the 924 in the U.S., Schutz hints broadly that it might do so altogether eventually. "Our future lies in building unique, high-technology Porsche components, not assembling produced items." That's clearly a reference to the 924's Audi engine, which Schutz openly admits was its weakest point, and almost in the same breath he just as clearly denies the current rumor in Europe that Porsche is collaborating with Volkswagen on a less expensive, front-drive sports car. The emphasis will be on the upper end of the sports car market.

And, adds Schutz, on increasing performance significantly while improving efficiency a little at a time: "We think

---

*Louise Porsche, sister of Ferry Porsche, married Dr Anton Piëch. Today, although the firm is now a stock company, the two families still guide its activities.

SCHUTZ

that's what our customers want," he explains. "In developing more efficient engines, you have a choice in marketing. You can take all your improvements out in efficiency, all in performance, or some in both areas. The 944 is a good example: much better performance, marginally better economy."

The 928 will also be the recipient of new, 944-style technology, especially with respect to combustion. "The 944 already has a high compression ratio, and 12:1 is definitely in the cards," Schutz adds; experimentation with a "thermodynamically optimized" engine is paving the way for this improvement to the 928's big V-8 engine, and a cylinder-cutout system is also a virtual certainty for the future. Yet another efficiency—and performance—boost for the 928 is likely to come in the form of a mechanical gearbox that is shifted, manually or automatically, under power. Schutz notes that this is also ideal for turbocharged engines for racing or the road.

Schutz's openness surfaces again in further discussion of the 928, whose sales fell 10 percent worldwide from 1980 to 1981 but picked up in the U.S. When asked about a German magazine's recent long-term test that showed the 928 to be less than remarkably reliable, on the one hand Schutz grumbled, "Anyone who tears up a clutch in 15,000 kilometers needs driving lessons," and on the other added that the 928 "is a complicated car for which we needed a few years to build like one expects a Porsche to be built. We must put its manufacturing quality where its technology is."

At the bottom end of Porsche's price scale, the Japanese cars' challenge has already been felt painfully in Stuttgart; in a recent interview with a Swiss publication, Schutz called them "competition to take seriously," added that they build their cars very well indeed and admitted that they offer more value for the money. He also declines to bury his head in the sand to the possibility that the Japanese might eventually produce competitors for the upper-class Porsches. "We'll compete with them as we best know how: by staying attuned to our customers, by emphasizing technology and quality," he says.

At the same time, Peter Schutz's American attitude comes through on the question of the 944 and its price. He wants to emphasize the top of the line and perhaps steer clear of the Japanese; yet thanks partly to the current low value of the Deutsche Mark and his competitive spirit, Porsche has set a relatively attractive price for the 944.

And how does Peter Schutz feel about racing? "Racing does three things for us," he answers. "The public-relations impact is obvious. It's also a good way to try things out—turbochargers, brakes, tire-pressure sensing devices, antiskid systems and so forth. As recently as six weeks before the 1981 Le Mans race we didn't have a head gasket for the 944 engine that would hold. After the race we did. Racing also plays an important role in motivating our people; we're sure that the opportunity to work on a race car is part of what Porsche is all about."

Obviously, racing will continue to be a big part of what Porsche is all about, and given Schutz's inclination to include women in his considerations, there could be some Porsche support for a female racing team in the not too distant future.

Schutz's leadership of Porsche can be expected to take not only the woman driver, but also the American market more closely into consideration than before—and this has not been a Porsche weakness in the past, to be sure. But can the leadership style of an American manager without long experience at the altar of European tradition be good for Porsche in the long run?

That's a premature question. Things are picking up at Porsche, even though the European industry is still in the doldrums. Though production for the 1980–1981 Porsche fiscal year fell 12 percent behind the 1979–1980 period, Porsche delivered 31,500 cars in calendar 1981, almost as many as in calendar 1980. Stocks have been reduced, Stuttgart (where 911s and 928s are built) is operating at capacity and Neckarsulm is up 20 percent from a year ago. But to attribute any of this to Schutz would surely be jumping the gun.

Schutz himself is modest about his assignment here in the equally modest Porsche administration building in the Stuttgart suburb of Zuffenhausen. "My job is to formulate the company's strategic plans and translate them into concrete actions. To recruit the best people and create a climate where people find personal fulfillment. It's my job to set the direction: If we build cars that don't fit the requirements of our customers and don't motivate our people, we won't get the results."

Manfred Jantke, who wasn't present during the interview, confirms it independently the next day: "He is strictly result-oriented." One has to note that Peter Schutz is also market-oriented. It will be fascinating to see how his approach to guiding Porsche—a stark contrast to the more purely technical approach of Dr Fuhrmann—affects Porsche's cars of the future.

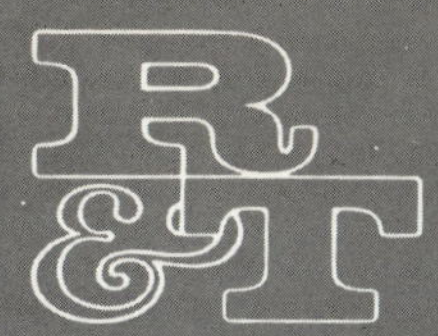

ROAD&TRACK
ON PORSCHE 1975-1978